The Luteal Phase

WILEY SERIES ON
CURRENT TOPICS IN REPRODUCTIVE ENDOCRINOLOGY

Series Editor
PROFESSOR S.L. JEFFCOATE
Chelsea Hospital for Women, London

Volume 1: Androgens and Anti-androgen Therapy

Volume 2: Progress Towards a Male Contraceptive

Volume 3: Ovulation: Methods for its Prediction and Detection

Volume 4: The Luteal Phase

CURRENT TOPICS IN REPRODUCTIVE
ENDOCRINOLOGY

Volume 4

The Luteal Phase

edited by

S.L. Jeffcoate
Chelsea Hospital for Women, London

A Wiley Medical Publication

JOHN WILEY & SONS
Chichester · New York · Brisbane · Toronto · Singapore

Copyright © 1985 by John Wiley & Sons Ltd.

All rights reserved.

No part of this book may be reproduced by any means, nor transmitted, nor translated into a machine language without the written permission of the publisher

Library of Congress Cataloging in Publication Data:
Main entry under title:

The luteal phase.

   (Current topics in reproductive endocrinology; v. 4)
(A Wiley medical publication)
   Includes index.
   1. Corpus luteum. I. Jeffcoate, S.L. II. Series. III. Series: Wiley medical publication. [DNLM: 1. Corpus Luteum – physiology – congresses.
W3 WI492 v. 4 / WP 520 1973 1983]
QP261.L88   1985   612'.62   84-17424
ISBN 0 471 90612 3

British Library Cataloguing in Publication Data:
The Luteal phase. – (Wiley series on current topics in
   reproductive endocrinology; v.4)
   1. Corpus luteum
   I. Jeffcoate, S.L.
   612'.62   QP261
ISBN 0 471 90612 3

Typeset by MHL Typesetting Ltd, Coventry
Printed in Great Britain by Page Bros. (Norwich) Ltd.

Contents

Series Preface vii

Preface ix

Chapter 1   Cellular Aspects of Corpus Luteum Function 1
             Stephen G. Hillier and E. Jean Wickings

Chapter 2   Control of Luteolysis 25
             David T. Baird

Chapter 3   Uterine Responses to the Corpus Luteum 43
             C.A. Finn

Chapter 4   Interaction between the Embryo and the Corpus
             Luteum 61
             D.K. Edmonds

Chapter 5   Prolactin and the Corpus Luteum 71
             Alan S. McNeilly

Chapter 6   Luteal Function after Ovulation Induction by Pulsatile
             Luteinizing Hormone Releasing Hormone 89
             *S. Franks, Z. van der Spuy, W.P. Mason, J. Adams
             and H.S. Jacobs*

Chapter 7   The Abnormal Luteal Phase 101
             J.R.T. Coutts

Chapter 8   Contraception in the Luteal Phase 115
             M.G. Elder

Index 123

Series Preface

Human reproductive endocrinology is a field that has seen great advances in knowledge in recent years. This has been based initially upon refinements in laboratory techniques which have led to an expansion of our understanding of the biochemical and physiological mechanisms underlying the control of reproduction in man and other animals, and subsequently to improvements in the diagnosis and treatment of patients.

This series of books is based upon a series of annual one-day symposia held at the Institute of Obstetrics and Gynaecology in the University of London. The aim is to cover an individual topic within the broad subject of reproductive endocrinology, giving the experimental background where appropriate, and to detail the current application to human endocrinology and clinical medicine.

September, 1981   S.L. JEFFCOATE

Preface

The luteal phase is the time after ovulation, when the steroid hormones from the corpus luteum prepare the uterine endometrium to receive and nurture a blastocyst. This volume thus carries on from the point where the previous volume in the series — on follicular development and ovulation — ended.

This three-way interaction — between the corpus luteum, endometrium and blastocyst — is still incompletely understood but it is clear that its proper functioning is necessary for conception in the human. Abnormalities in the luteal phase can thus be a potentially treatable cause of infertility and the normal luteal phase can be a site for contraceptive action.

In this volume, as in previous books in the series, the physiological and biochemical background is covered in the early chapters. Later chapters deal with abnormalities of the luteal phase and infertility; finally, current approaches to contraception in the luteal phase are described.

This volume should be of interest to clinicians and scientists working in the field of reproduction including those involved in the management of the infertile couple.

August, 1984   S.L. JEFFCOATE

The Luteal Phase
Edited by S.L. Jeffcoate
© 1985 John Wiley & Sons Ltd.

CHAPTER 1

Cellular aspects of corpus luteum function

STEPHEN G. HILLIER AND E. JEAN WICKINGS

Institute of Obstetrics and Gynaecology,
Hammersmith Hospital,
DuCane Road,
London W12 0HS, UK

Introduction

The luteal phase of the human menstrual cycle reflects the functional lifespan of the corpus luteum (Baird *et al.*, 1975). Progesterone is its principal secretion although it also produces oestrogen: as much if not more than does the preovulatory follicle during the follicular phase of the cycle (Young, 1961). The cellular mechanisms underlying the secretion of both steroids by the human corpus luteum are discussed here.

Although the beginning of the luteal phase is usually defined as the moment of ovulation, it actually starts with the onset of the mid-cycle luteinizing hormone (LH) surge which precedes follicle rupture by some 24 hours or more. This is when structural and biochemical changes heralding the onset of luteinization are initiated in the preovulatory follicle (Edwards *et al.*, 1980; World Health Organization, 1981). It is logical and convenient to consider the cellular basis of corpus luteum function in three progressive stages defined naturally by the changes which (1) occur in the preovulatory follicle leading up to the LH surge, (2) are induced in that follicle during the surge and which culminate in ovulation, and (3) occur after ovulation as the corpus luteum proper forms (Ross and Hillier, 1978). These stages are orientated to the normal menstrual cycle pattern of plasma hormone levels in Figure 1.

1. The Preovulatory Follicle (Before the LH Surge)

Two or three days before the onset of the LH surge, the preovulatory follicle

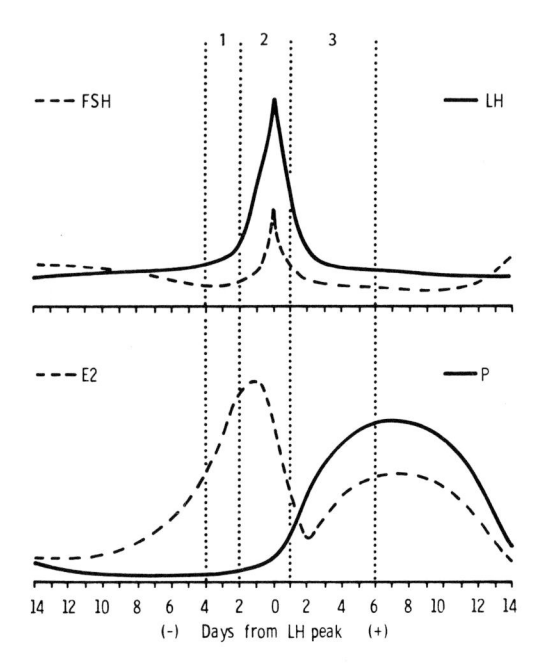

Figure 1. Stages of corpus luteum formation orientated to the normal menstrual cycle pattern of plasma hormone levels: (1) preovulatory follicle (before the LH surge); (2) ovulatory follicle (during the LH surge); (3) corpus luteum proper (after ovulation)

will have emerged as the largest healthy follicle in either ovary (Gougeon and Lefevre, 1983). Its rate of growth is still linear due to increasing granulosa cell numbers and accumulation of follicular fluid. By the time the surge starts, it will be more than 20 mm in diameter (Kerin et al., 1981; McNatty, 1982). At this time it is characterized biochemically primarily by its tremendous rate of oestradiol secretion, and, as the almost exclusive source of the oestradiol circulating in plasma, it can secrete up to about 400 μg oestradiol per day (Baird and Fraser, 1974).

Observations *in vitro* of preovulatory granulosa cells and follicular fluid have shown that in the dominant follicle the activity of the key enzyme involved in oestradiol synthesis, aromatase, is 10–100 times higher than in non-ovulatory follicles present in either ovary (Hillier et al., 1981; McNatty et al., 1983). This high activity is reflected in a proportionately high follicular fluid oestradiol level (Figure 2).

The low plasma level of progesterone which is a feature of this stage of the menstrual cycle (Figure 1) belies the situation with respect to progesterone synthesis in the preovulatory follicle. The requisite enzyme systems have also become established but they require full activation by the impending LH surge (Channing, 1980; Hillier et al., 1983). This latent potential for progesterone

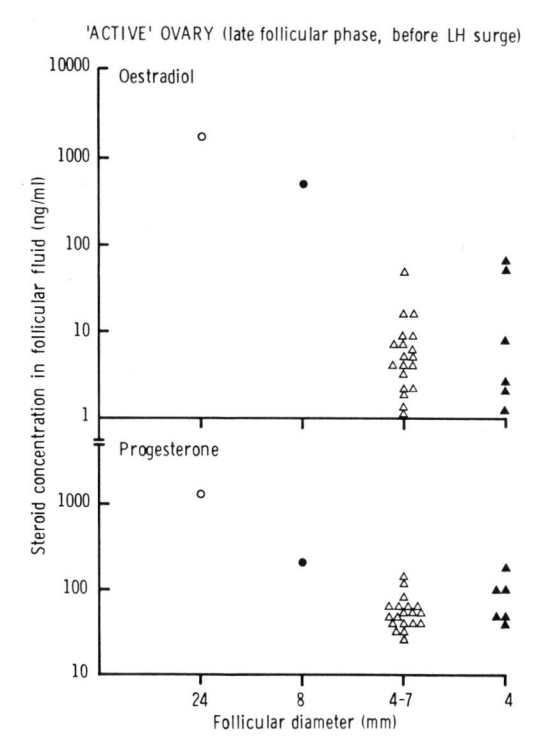

'ACTIVE' OVARY (late follicular phase, before LH surge)

Figure 2. Concentration of oestradiol (*upper panel*) and of progesterone (*lower panel*) in follicular fluid from the individual follicles present in a human ovary shortly before onset of the LH surge. O, 24-mm diameter preovulatory follicle; ●, 8-mm diameter follicle; △, 4–7-mm diameter follicles; ▲, 4-mm diameter follicles. The ovary was resected during laparotomy on day 12 of a (usually) 28-day menstrual cycle. (This is patient No. 10 in the series described by McNatty *et al.*, 1983)

secretion is suggested from an examination of follicular fluid steroid levels (Figure 2). In the dominant follicle, the progesterone level often exceeds 1 µg/ml and is invariably an order of magnitude or more higher than in the non-ovulatory follicles that are present (McNatty, 1978).

The granulosa cell layer is the principal site of oestradiol and progesterone formation in the preovulatory follicle (McNatty *et al.*, 1979; Hillier, 1981; and see below). Granulosa cell enzyme systems required for the synthesis of both steroids are under primary follicle-stimulating hormone (FSH) control during early preovulatory development (Dorrington *et al.*, 1983; Hsueh *et al.*, 1983) but become responsive to LH later on. Important correlates of FSH action during granulosa cell differentiation are listed in Table 1. This list includes LH receptor induction (Zeleznik *et al.*, 1974). Once the granulosa cell LH receptor system becomes expressed following appropriate stimulation of the cells by

Table 1. Key granulosa cell functions induced by FSH

- Follicular fluid formation
  (secretion of proteoglycans)
- Aromatization of androgens
  (oestrogen biosynthesis)
- Steroidogenesis
  (progesterone biosynthesis)
- LH and prolactin responsivity
  (receptor induction)

FSH and oestradiol (Richards, 1980), biochemical events initially under exclusive FSH control become directly responsive to LH also (Channing *et al.*, 1978; Hillier *et al.*, 1978). This includes oestradiol and progesterone synthesis. In the mature preovulatory follicle (and subsequently the corpus luteum; see below) each gonadotrophin can interact with its corresponding granulosa cell receptor system to control these activities via a common cyclic AMP dependent subcellular mechanism (Birnbaumer and Kirchik, 1983).

Preovulatory granulosa cell aromatase activity reaches a maximum before the onset of the LH surge when it becomes refractory to further stimulation by FSH and LH *in vitro* (Hillier *et al.*, 1983) (Figure 3). However, sustained expression of maximal activity *in vivo* may depend upon appropriate stimulation by LH at this time (see below).

In contrast, measurements made *in vitro* before the LH surge show that preovulatory granulosa cell progesterone synthesis is minimal (Channing, 1980; Channing and Reichert, 1983; Hillier *et al.*, 1983) (Figure 3). However, the addition of FSH or LH to the incubation medium can elicit striking increases in progesterone production which become readily apparent within 3 h (Hillier *et al.*, 1983) (Figure 3).

Thus, before the onset of the LH surge the situation with regard to subsequent corpus luteum function can be summarized as follows: (i) oestrogen secretion by the preovulatory follicle approaches a maximum and is reflected in the activity of its granulosa cell aromatase system and high follicular fluid oestradiol level; (ii) progesterone secretion is still minimal although follicular fluid from the preovulatory follicle contains the highest level of progesterone suggesting that active synthesis is under way; (iii) preovulatory granulosa cells have acquired the capacity to respond to FSH and LH with a rapidly increased rate of progesterone production.

2. The Ovulatory Follicle (During the LH Surge)

The mid-cycle LH surge starts at or around the time that the preovulatory follicle achieves a maximal rate of oestradiol secretion (World Health Organization,

GRANULOSA CELLS FROM PREOVULATORY FOLLICLES
(before LH surge; n = 4)

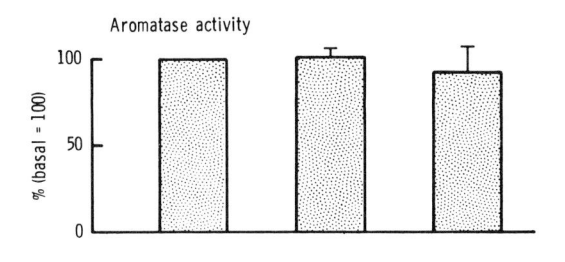

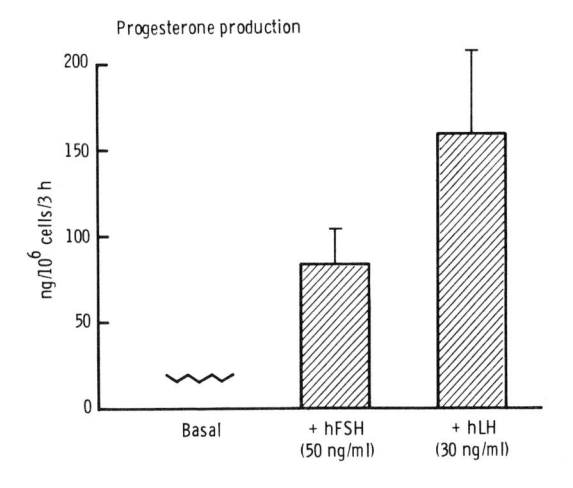

Figure 3. Human granulosa cell aromatase activity (*upper panel*) and progesterone production (*lower panel*) during short-term (3 h) incubation *in vitro*. Cells aspirated from the preovulatory follicle ($\geqslant 19$ mm diameter) in four tubal-surgery patients who underwent laparotomy shortly before onset of the mid-cycle LH surge were studied (Hillier et al., 1983). Granulosa cell suspensions (approximately 0.5×10^4 cells in 0.5 ml of medium 199 containing 0.1% bovine serum albumin) were incubated at 37°C in the presence or absence of FSH (hFSH LER 8/116) or LH (hLH LER 960) as indicated. Aromatase activity was determined as oestradiol formation from exogenous precursor (10^{-7} M testosterone) (Hillier et al., 1981, 1983). Basal activity (no added FSH or LH present) was taken as 100%. Progesterone production was assessed as steroid accumulated in the medium in the absence of aromatase substrate. All steroid measurements were by radioimmunoassay. Results are the mean ($\pm$ SEM) from all four follicles. (Three of these follicles are described in the study by Hillier et al., 1983)

1980; Testart *et al.*, 1982b; Hoff *et al.*, 1983) (Figure 1). A simultaneous increase occurs in the plasma progesterone level reflecting the onset of luteinization in response to LH (Djahanbakhch *et al.*, 1983). As this early progesterone rise gains momentum there is a sudden precipitate drop in the plasma oestradiol level which usually occurs 24 h or so before follicular rupture (Hoff *et al.*, 1983). Within the follicle, a chain of biochemical events is triggered which culminates in ovulation approximately 37 h after onset of the LH surge (Edwards *et al.*, 1980) (Table 2). Changes in follicular fluid steroid levels parallel those occurring

Table 2. Intrafollicular events triggered by the LH surge

- Changes in steroid biosynthesis
- Prostaglandin biosynthesis
- Resumption of oocyte meiotic maturation
- Cumulus expansion
- Reduced cumulus cell–oocyte coupling
- Synthesis of preovulatory enzymes

in plasma; oestradiol falls as progesterone rises (Edwards *et al.*, 1980; Testart *et al.*, 1982a) (Figure 4). A fall in follicular fluid androgen levels is also noted; this is believed to be directly related to the fall in oestradiol (Edwards *et al.*, 1980) (see below).

The follicular cell functions underlying these dramatic changes are as follows.

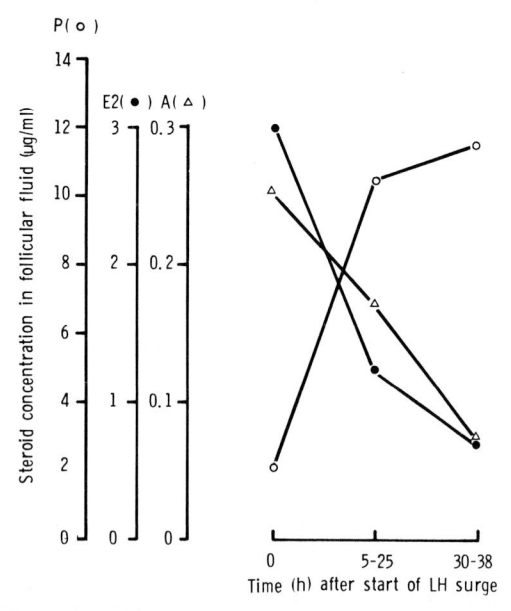

Figure 4. Changes in human follicular fluid progesterone (P,○), oestradiol (E2, ●) and androstenedione (A, △) levels in the ovulatory follicle after onset of the LH surge. (Reproduced from Testart *et al.*, 1982a, with permission)

2.1. Progesterone Biosynthesis

Luteinizing granulosa cells are the primary source of progesterone secreted by the ovulatory follicle (Channing, 1980; McNatty et al., 1979). The follicle studied in Figure 5 had grown to a diameter of approximately 28 mm before it was dissected from a human ovary approximately one day after onset of the LH surge (Hillier et al., 1981). Granulosa cells and theca cells were separated and maintained in tissue culture to determine relative abilities to produce progesterone *in vitro*. Although theca cells undertook some production and responded well to human chorionic gonadotrophin (hCG; used as a surrogate for LH), granulosa cells produced several hundred times more steroid on a cell-for-cell basis. That they were also refractory to hCG suggests that maximal activation had already been achieved during exposure to the surge level of circulating LH *in vivo*.

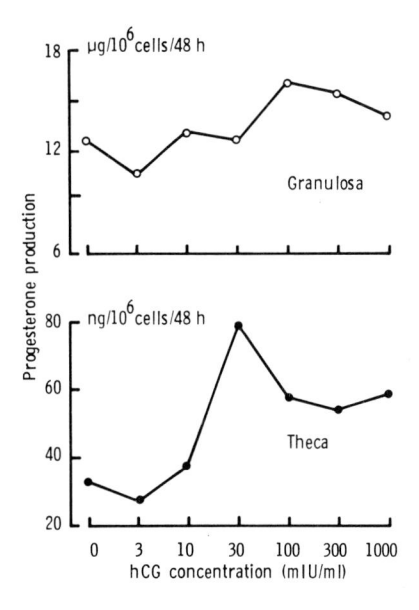

Figure 5. Progesterone production by cultured granulosa cells (*upper panel*) and theca cells (*lower panel*) isolated from a 28-mm diameter human ovulatory follicle approximately 12–24 h after onset of the mid-cycle LH surge. (This was tubal surgery patient P in the series described by Hillier et al., 1981.) Monolayer cell cultures were established at the same cell density (2 × 10^5 cells per 2 ml of medium 199 containing 5% donor calf serum); the theca layer was dispersed with 0.05% collagenase. Incubation was for 48 h at 37°C in a humidified incubator gassed with 5% CO_2 in air. Increasing concentrations of hCG (CR 119) were present in the culture medium, as indicated. Progesterone accumulation in culture medium was measured by radioimmunoassay: note the different scales used for the two cell types. Each datum point is the mean of an incubation in triplicate

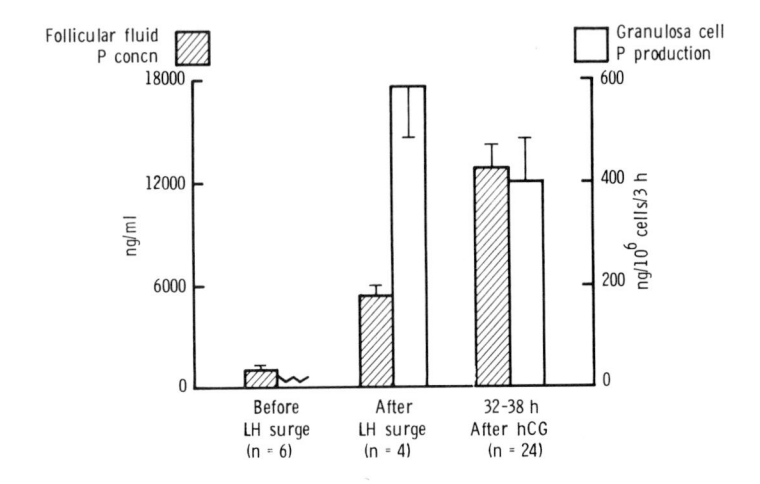

Figure 6. Progesterone (P) concentration in follicular fluid and progesterone production *in vitro* by granulosa cells aspirated from human ovulatory follicles before and approximately 12–24 h after onset of a spontaneous LH surge (data from Hillier *et al.*, 1981, 1983), and 32–38 h after injection of hCG to induce ovulation in patients undergoing *in vitro* fertilization (Hillier *et al.*, 1984). Progesterone production by washed granulosa cells was assessed as steroid accumulating in the medium over a 3-h incubation at 37°C (see the legend to Fig. 3 for details) corrected for endogenous steroid levels present in unincubated (frozen) cell suspensions. Results are the mean ($\pm$ SEM) of the number of observations indicated

Examination *in vitro* of granulosa cell progesterone production in relation to follicular fluid progesterone concentration reveals the pattern shown in Figure 6. Both parameters increase, if not entirely proportionately, as the follicle approaches ovulation. It can be calculated from the data shown that between the time the LH surge begins (or that hCG is injected) and the time of follicular rupture the progesterone-producing capacity of luteinizing granulosa cells increases more than one hundred fold. There can be no doubt that these cells are the major source of the rising peripheral progesterone level which occurs between onset of the LH surge and ovulation.

2.2. Oestradiol Biosynthesis

More than 90% of the preovulatory follicle's capacity for aromatization is located in its granulosa cells (Hillier *et al.*, 1981). Yet since granulosa cells are unable to synthesize androgen (required as an aromatase substrate) they depend upon an extracellular source of substrate for use as the oestrogen precursor (Armstrong *et al.*, 1978). Theca cells responding to stimulation by LH are believed to be the primary source of the androgen used by granulosa cells for this purpose (Baird, 1977a; Moor, 1977; Tsang *et al.*, 1979). It was pointed out

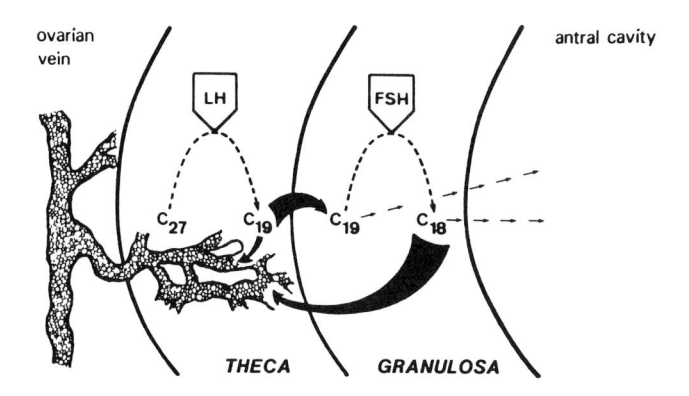

Figure 7. Current concept of gonadotrophin-controlled cellular interactions which co-ordinate follicular oestrogen biosynthesis in the human preovulatory follicle. Androgen (C_{19}) is synthesized from cholesterol (C_{27}) in the vascularized theca interna due to stimulation by LH. The avascular granulosa cell layer stimulated by FSH has an active androgen aromatase enzyme system. Thecal androgen which diffuses across the lamina basalis serves as the aromatase substrate. Cells in the parietal granulosa layer may be the most active sites of aromatization and are adjacent to the extensive network of blood vessels which encapsulates the lamina basalis. The oestrogen (C_{18}) they produce diffuses back into the blood stream. Some steroid, androgen and oestrogen also passes into the antral cavity. Before onset of the LH surge, the follicular fluid level of oestrogen (absolute and relative to androgen) increases in proportion to the activity of the granulosa cell aromatase activity. However, after surge onset this relationship breaks down, as discussed in the text. (Reproduced from Hillier, 1981, with permission)

earlier (see Table 1) that the initial induction of the granulosa cell aromatase system depends upon appropriate stimulation by FSH (Dorrington et al., 1983). Thus the two follicular cell types and both gonadotrophins are thought to interact in the control of follicular oestrogen synthesis, as illustrated in Figure 7 (Ryan, 1979).

Constant stimulation by FSH and/or LH may be essential for luteinizing granulosa cells to retain a fully activated aromatase system (Hillier et al., 1983). This is suggested by observations in vitro of granulosa cells recovered from the human ovulatory follicle after the onset of the LH surge (Figure 8). In the example shown, granulosa cell aromatase activity was measured immediately after follicle aspiration and repeated 48 h later on cultured cell monolayers derived from the same aspirate. When incubated in the absence of exogenous gonadotrophins, the activity fell to a fraction (approximately 25%) of that observed for freshly isolated cells. However, in the continued presence of FSH or LH this loss was substantially ameliorated.

The relationship between aromatase activity in vitro and the follicular fluid oestradiol level prevailing at the time of granulosa cell isolation is illustrated in

GONADOTROPHIC MAINTENANCE OF GRANULOSA
CELL AROMATASE ACTIVITY DURING CULTURE
(ovulatory follicle: LH surge started)

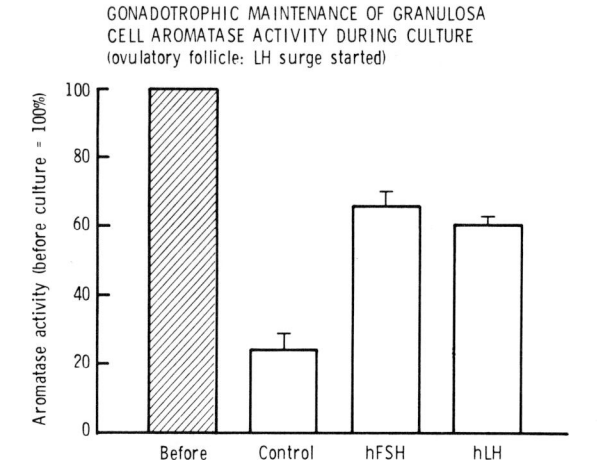

Figure 8. Granulosa cells recovered from an ovulatory human follicle 12–24 h after onset of the mid-cycle LH surge (patient XXIX in the series described by Hillier *et al.*, 1983): effect of culture for 48 h in the absence (control) or presence of FSH (75 ng of hFSH LER 8/116 per ml) or LH (30 ng of hLH LER 960 per ml) on aromatase activity. Aromatase was assessed by incubating the freshly aspirated cells (0.5×10^4 cells per 0.5 ml medium 199 containing 0.1% BSA) for 3 h in the presence and absence of 10^{-7} M testosterone and measuring oestradiol formation from the substrate by radio-immunoassay. The assay was repeated 48 h later on the culture cell monolayers. All incubations were in quadruplicate. Results are expressed as mean percentage activity ($\pm$SEM) relative to that of the freshly aspirated cells

Figure 9. The fall in follicular fluid oestradiol concentration which occurs as the LH/hCG-stimulated follicle approaches ovulation is not accompanied by a commensurate fall in the granulosa cell potential for synthesis of the steroid. Instead the supply of thecal aromatizable androgen is reduced following the LH surge (Moor, 1974; Leung and Armstrong, 1980). Thus the fall in follicular fluid oestradiol concentration is paralleled by a similar fall in androstenedione, quantitatively the most important androgen produced by the theca of the human follicle (McNatty *et al.*, 1979; Tsang *et al.*, 1979). It has yet to be explained how the mid-cycle LH surge acts to modify thecal androgen synthesis in this way. Possibly a 'desensitization' response is elicited by the gonadotrophin leading to lesions in the pathways of androgen formation (McNatty *et al.*, 1980; Fritz and Speroff, 1982).

It might seem paradoxical that thecal androgen formation is suppressed by surge LH levels whereas the granulosa cell aromatase system is not only retained but may even depend upon LH support before the follicle ovulates. How-

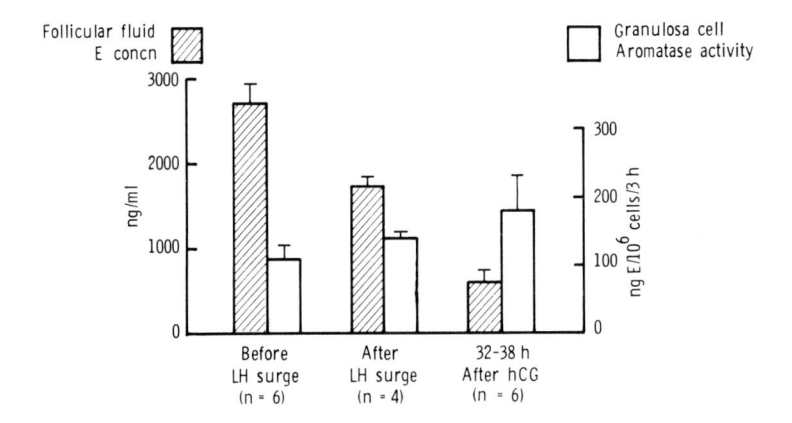

Figure 9. Oestradiol (E) concentration in follicular fluid and aromatase activity in vitro of granulosa cells aspirated from human ovulatory follicles before and approximately 12–24 h after onset of a spontaneous LH surge (data from Hillier et al., 1981, 1983), and 32–38 h after injection of hCG to induce ovulation in patients undergoing in vitro fertilization (Hillier et al., 1984). The aromatase activity of washed cell suspensions was determined as described in the legends to Figs. 3 and 8. Measurements of oestradiol were carried out by radioimmunoassay and the results are the mean (±SEM) of the number of observations indicated

ever, as discussed below, the retention of active aromatase by granulosa-lutein cells might be integral to the resumption of oestrogen secretion which the corpus luteum eventually undertakes.

Thus essential features of the interval between the onset of the LH surge and ovulation are: (i) the potential for granulosa cell progesterone secretion is activated by LH in the ovulatory follicle; (ii) the potential for granulosa cell oestrogen synthesis (aromatase activity) is not apparently altered by LH although follicular oestrogen secretion actually declines due at least in part to a diminished supply of aromatizable androgen by the theca to the granulosa cells; (iii) maintenance of an active aromatase system may depend upon appropriate gonadotrophic support right up until ovulation and this is likely to be essential as the newly formed corpus luteum gradually resumes the task of oestradiol secretion (see below).

3. The Corpus Luteum (After Ovulation)

If the ovulatory follicle received the appropriate combination and degree of gonadotrophic support during the antecedent follicular phase, two factors will largely dictate how it functions as the corpus luteum. These are its blood supply and the luteotrophic support it receives from the pituitary gland.

3.1. Vascular Control of Corpus Luteum Function

With follicular rupture and extrusion of the oocyte, the vasculature previously confined to the theca layer extends into the granulosa (Harrison and Weir, 1977). Production by granulosa-lutein cells of an angiogenic factor is believed to be involved (Gospodarowicz and Thakral, 1978). Blood flow to the corpus luteum increases to a maximum at or about the mid-luteal phase of the cycle when the vasculature is fully developed (Reynolds, 1973). As this occurs the process of luteinization, morphological and functional, gains momentum. It has been estimated that when the plasma progesterone level is maximal during the mid-luteal phase (Figure 1) the corpus luteum secretes more than 25 mg of progesterone per day (Baird, 1977b). Since this rate of production is too high to be sustained by steroid synthesized exclusively from endogenous precursor cholesterol, an exogenous source must also be used. This is believed to be in the form of blood-borne lipoprotein cholesterol (Simpson et al., 1980). Granulosa-lutein cells like many other cells can derive unesterified cholesterol by uptake and degradation of low density lipoprotein (LDL) (Goldstein and Brown, 1977; Tureck and Strauss, 1982). The uptake can be stimulated by hCG and requires interaction of LDL with specific receptors in the cell membrane (Carr et al., 1981; Ohashi et al., 1982). The biochemistry of LDL uptake and hydrolysis of cholesterol esters has recently been reviewed (Gwynne and Strauss, 1982).

Carr et al. (1982) have suggested how neo-vascularization of the granulosa cell layer might be a primary factor in the initiation and maintenance of corpus luteum progesterone secretion (Figure 10). Thus, before ovulation when the lamina basalis is still intact, granulosa cells are not exposed directly to blood although they are in contact with serum-derived factors in follicular fluid (Edwards, 1974). Measurements of serum protein levels in follicular fluid suggest the existence of a 'blood–follicle barrier' (Shalgi et al., 1973) (Figure 11). It seems to behave as a molecular sieve which facilitates passage of protein in inverse proportion to molecular weight and/or size. High density lipoprotein (HDL), a relatively small molecule, is found in human preovulatory follicular fluid at a level similar to that in blood (Simpson et al., 1980; unpublished data of K.P. McNatty and S.G. Hillier) (Table 3). However, compared to LDL, HDL has been shown to be an extremely poor source of precursor cholesterol for human granulosa cells (Tureck and Strauss, 1982; Carr et al., 1981). Conversely, LDL, from which these cells can derive precursor cholesterol, is almost completely excluded from follicular fluid since it is such a large molecule (Shalgi et al., 1973; Simpson et al., 1980) (Figure 11 and Table 3). Therefore although LH-activated granulosa cells possess the potential for progesterone synthesis (see above), one reason that it is not fulfilled until ovulation might be that precursor cholesterol derived from LDL is rate-limiting. When the lamina basalis is breached and blood vessels invade the granulosa layer, granulosa-lutein cells become exposed to blood-borne LDL cholesterol and undertake the high rate of progesterone synthesis characteristic of the postovulatory period. The

CELLULAR ASPECTS OF CORPUS LUTEUM FUNCTION 13

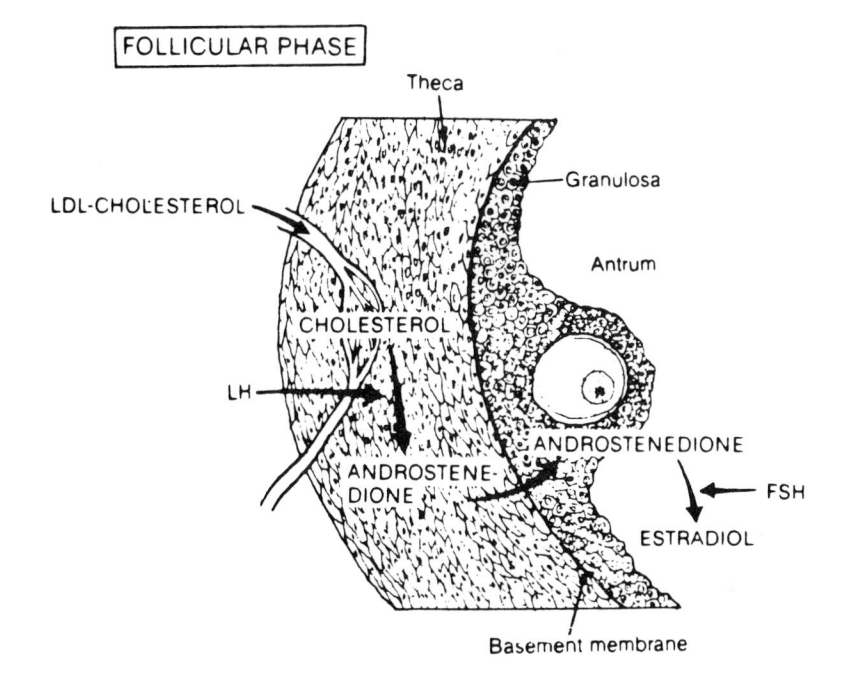

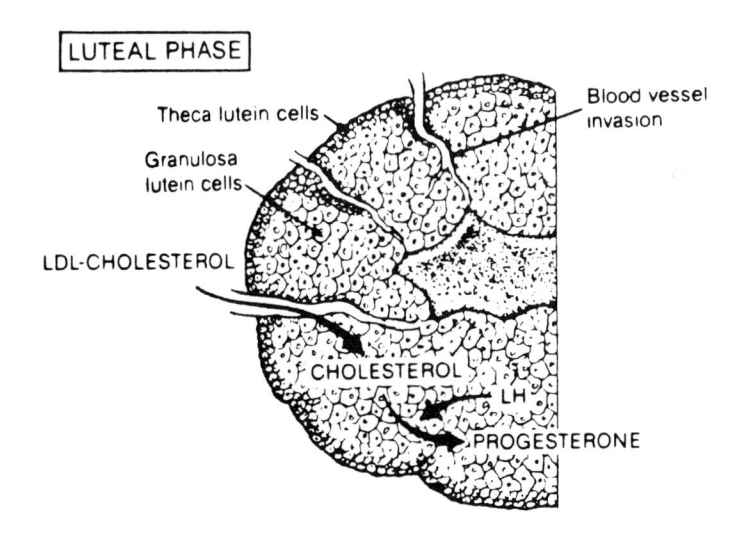

Figure 10. Model proposed for the regulation of progesterone secretion by the human ovary. Follicular phase (*top*), luteal phase (*bottom*). (Reproduced from Carr *et al.*, 1982, with permission)

THE BLOOD-FOLLICLE BARRIER

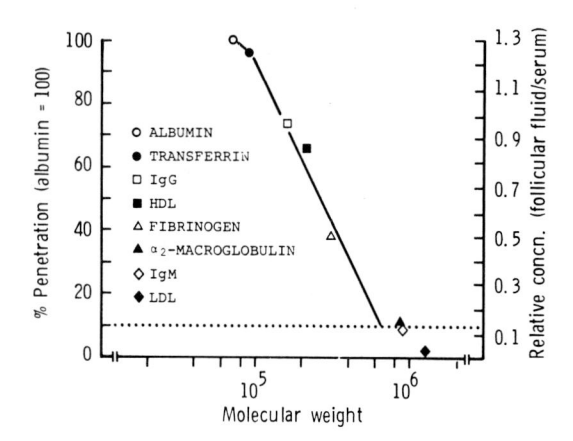

Figure 11. The blood–follicle barrier (redrawn from Shalgi et al., 1973, with permission). The datum point for HDL is the mean of five measurements on individual preovulatory human follicular fluids and corresponding serum samples carried out in the author's laboratory (unpublished data from K.P. McNatty and S.G. Hillier). All the other measurements are those originally published by Shalgi et al. (1973)

newly developing blood supply allows rapid drainage of the corpus luteum and dispersal of the steroid. Maximal capillarization is achieved about one week after ovulation, which corresponds with the time of maximal progesterone secretion by the corpus luteum (Baird et al., 1975; Bjersing, 1981). Measurements of LDL in blood-contaminated fluid aspirated from the core of the fully formed corpus luteum and from luteinized unruptured follicles confirm the presence of LDL at levels approaching those in serum (Table 3).

3.2. Gonadotrophic Support of Corpus Luteum Function

For the maintenance of progesterone secretion and the resumption of oestradiol secretion, the corpus luteum requires gonadotrophic support (di Zerega and Hodgen, 1980; Hutchison and Zeleznik, 1983). An opportunity to evaluate the gonadotrophic control of granulosa-lutein cell steroidogenesis is afforded by the availability of follicular aspirates taken shortly before ovulation during the collection of oocytes for fertilization *in vitro* (Edwards et al., 1980). Multiple ovulatory follicles are usually induced in these patients by treatment with clomiphene citrate and/or human menopausal gonadotrophin (Hillier et al., 1984). We have separated granulosa-lutein cells from such aspirates and studied them for up to six days (i.e. equivalent to almost half the luteal phase) as monolayer cell cultures (Hillier et al., 1983). As few as 1000 cells are adequate to

Table 3. Lipoprotein levels (mg per 100 ml) in human follicular fluid (FF) and plasma (PL). (Unpublished data from K.P. McNatty and S.G. Hillier)

Fluid source	HDL* FF	HDL* PL	LDL† FF	LDL† PL
Ovulatory follicle‡	370	334	<0.5	405
Ovulatory follicle§	330	448	<0.5	600
Corpus luteum¶	318	422	386	425
Luteinized unruptured follicle♭	246	458	63	349

* Determined by radial immunodiffusion.
† Determined by rocket immunoelectrophoresis.
‡ Aspirated approximately 12–24 h after onset of LH surge on day 12 of a normally 28-day menstrual cycle; contained no visible blood (Patient XXXV studied by Hillier et al., 1983.)
§ Aspirated approximately 12–24 h after onset of LH surge on day 12 of a normally 28-day menstrual cycle; contained no visible blood. (Patient 18 studied by McNatty et al., 1983.)
¶ Aspirated from the core of a mid/late luteal phase corpus luteum on day 27 of a normally 28-day menstrual cycle; contained blood.
♭ Aspirated on day 18 of a normally 28-day menstrual cycle; contained blood. (Patient 9 studied by McNatty et al., 1983.)

establish a culture which retains functional and morphological integrity throughout the period of experimental observation *in vitro* (Figure 12).

Given an adequate source of exogenous precursor cholesterol in the form of lipoprotein (culture medium supplemented with serum), progesterone accumulates in granulosa-lutein cell cultures at an approximately linear rate over the first four days even in the absence of exogenous gonadotrophin (Figure 13). This could be explained by stimulation due to hCG associated with the cells at the time of isolation (Dennefors et al., 1983). However, progesterone accumulation is further stimulated by the addition of exogenous LH to the culture medium. The effect is usually most dramatic during the first two to three days of culture and is consistent with the widely accepted role of LH as the primary human luteotrophin (Marsh, 1976) (Figure 13). Occasional stimulation by FSH has also been noted (Figure 13). The FSH preparation (human FSH LER 8/116) used in these experiments was highly pure since it had been treated with α-chymotrypsin to reduce LH contamination selectively (Reichert, 1967). At the concentration used (30 ng/ml) there were approximately 27 mIU of FSH per ml but only approximately 0.15 mIU of LH per ml. Therefore a significant LH-effect at this dose seems unlikely although it cannot be completely excluded. Human corpus luteum cells possess measurable receptors for FSH during the early luteal phase of the cycle (McNeilly et al., 1980) so a role for the gonadotrophin in the control of early corpus luteum progesterone secretion is possible.

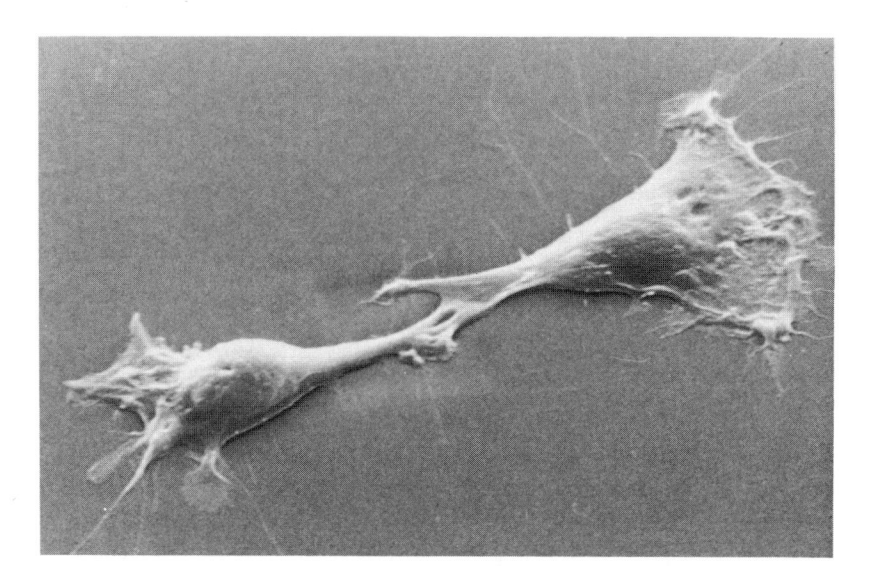

Figure 12. Scanning electron micrograph of human granulosa-lutein cells maintained in tissue culture for 6 days after aspiration from a follicle on the verge of ovulation. Steroid production by cells aspirated from the same follicle is shown in Figs. 13 and 14

Similar to preovulatory granulosa cells, granulosa cells from the periovulatory follicle are unable to synthesize oestradiol in the absence of exogenous precursor androgen (Bjersing, 1981). However, as noted above, a high level of aromatase activity is present in granulosa-lutein cells at the time of ovulation, and after aspiration it persists during tissue culture even in the absence of exogenous gonadotrophic support (Figure 14). However, the fact that the activity is further increased by the presence of FSH or LH in the culture medium suggests a role for either or both gonadotrophins in the regulation of this key enzyme system of the corpus luteum. Stimulatory effects of FSH on oestradiol production by human corpus luteum tissue have been described before (Hunter and Baker, 1981).

These new findings with isolated granulosa-lutein cells can be integrated with other information on the control of human luteal steroidogenesis as outlined in Figure 15. The scheme suggests that FSH acting through its own receptor system might affect steroidogenesis via the same intracellular cyclic AMP dependent mechanisms known to be under primary LH control (Birnbaumer and Kirchik, 1983). Although LH is probably the most important human luteotrophin (Marsh, 1976; Bjersing, 1981), the possibility that FSH plays a subordinate role, particularly during early corpus luteum function, warrants further attention.

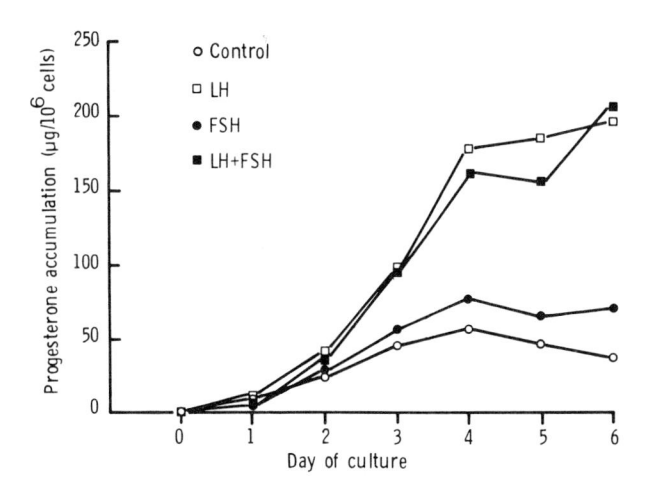

Figure 13. Progesterone accumulation in human granulosa-lutein cell cultures. The cells were aspirated from an ovulatory follicle in the ovary of a patient undergoing *in vitro* fertilization who was injected with 5000 IU of hCG 37 h earlier; although multiple follicles (due to clomiphene stimulation) were aspirated to obtain oocytes, only one aspirate was studied in this experiment (Hillier *et al.*, 1984). After separation from follicular fluid, the cells were dispersed with hyaluronidase, washed, counted and distributed into culture wells (Hillier *et al.*, 1983). The cultures (0.5 × 10^4 cells per 0.5 ml of medium 199 supplemented with antibiotics and 5% donor calf serum) were incubated for 1–6 days at 37°C in a humidified incubator gassed with 5% CO_2 in air. Sets of cultures (each in quadruplicate) were terminated on each day after incubating in the absence of exogenous gonadotrophin (O, control), or in the presence of LH (□, 30 ng/ml hLH LER 960), FSH (●, 30 ng/ml hFSH LER 8/116), or LH plus FSH (■). Progesterone in the culture medium was determined by radioimmunoassay. The mean level of steroid present on each day of culture is shown

If granulosa-lutein cells are a significant site of oestradiol synthesis in the human corpus luteum (Bjersing, 1981; Mori *et al.*, 1983), what is the source of the precursor androgen used as the aromatase substrate? Histological examination of the newly formed corpus luteum shows that the theca interna is no longer a discrete cell layer (World Health Organization, 1981). Perhaps this apparent dearth of theca-lutein cells reflects their demise with respect to androgen formation after exposure to the mid-cycle LH surge, as discussed above (Fujita *et al.*, 1981). However, as the newly formed corpus luteum develops and oestradiol secretion is re-established, the theca-lutein cells become more apparent as trabeculae among the enlarging granulosa-lutein cells and often in close proximity to the newly formed blood vessels (Van Look and Baird, 1980; World Health Organization, 1981). Human theca-lutein cells have been reported to undertake *de novo* androgen biosynthesis *in vitro* (Macnaughton *et al.*, 1981). The reactivation of this function in the corpus luteum might therefore

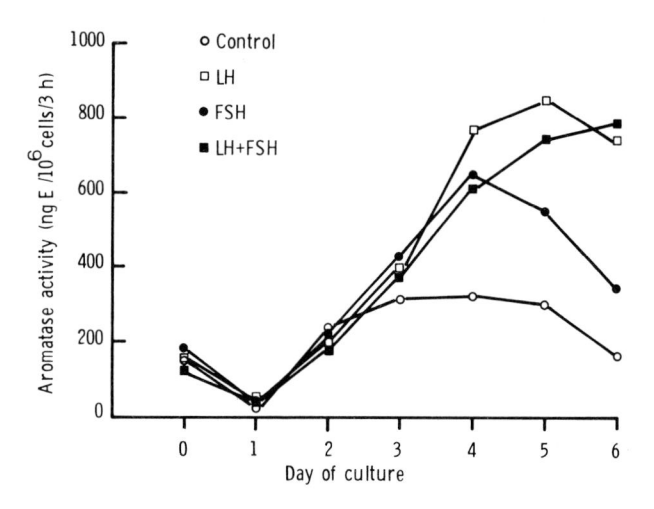

Figure 14. Aromatase activity in human granulosa-lutein cell cultures. (Details of the cell isolation and culture, and progesterone production by the same cultures are given in Fig. 13 and the corresponding legend.) Aromatase activity was measured in freshly isolated cells (day 0) as described in the legend to Fig. 8 except that precursor testosterone was added at 10^{-6} M. Daily measurement of aromatase in cell cultures was achieved as follows. Sets of quadruplicate cultures corresponding to each hormone treatment were terminated; before collecting the medium, testosterone (10^{-6} M) was added to two of the cultures and vehicle (5 µl of ethanol) was added to the other two. All cultures were incubated for a further 3 h at 37°C and the medium was collected to be measured for oestradiol (E) content by radioimmunoassay. The level of oestradiol produced in the presence of testosterone corrected for that which accumulated in its absence (minimal) provided the aromatase index. The mean result of each daily measurement is shown

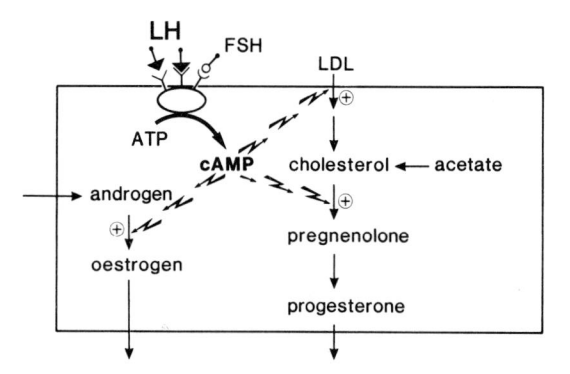

Figure 15. Gonadotrophic control of human granulosa-lutein cell steroidogenesis. LH, the primary human luteotrophin, acts through its membrane-associated receptor system to influence the cyclic AMP (cAMP) dependent steps in steroidogenesis illustrated (cf. March, 1976; Bjersing, 1981; Birnbaumer and Kirchik, 1983). It is suggested that FSH plays a subordinate role, particularly during early corpus luteum function, by affecting the same mechanisms via its own receptor system in the cell membrane

permit adjacent granulosa-lutein cells to recommence oestradiol secretion and give rise to the raised circulating level of oestradiol which characterizes the mid-luteal phase of the cycle (Figure 1) (Bjersing, 1981). The role of gonadotrophin in thecal reactivation has not been studied.

In summary, these studies indicate that corpus luteum function involves: (i) breakdown of the blood–follicle barrier and vascularization of the granulosa cell layer, ensuring the availability of LDL cholesterol for use as a precursor for progesterone synthesis in LH-activated granulosa-lutein cells; (ii) a major role for LH as the principal luteotrophin and a subordinate role for FSH, particularly in the newly formed corpus luteum; (iii) possible involvement of both gonadotrophins in the maintenance of granulosa-lutein cell aromatase activity after ovulation; (iv) reactivated theca-lutein cells as the most likely source of precursor androgen used by granulosa-lutein cells to synthesize oestradiol.

References

Armstrong, D.T., Goff, A.K., and Dorrington, J.H. (1978). 'Regulation of follicular estrogen biosynthesis.' In *Ovarian Follicular Development and Function* (eds. A.R. Midgley, Jr. and W.A. Sadler), pp. 169–181. Raven Press, New York.

Baird, D.T. (1977a). 'Evidence *in vivo* for the two-cell hypothesis of oestrogen biosynthesis in the sheep Graafian follicle.' *J. Reprod. Fertil.*, **50**, 183–185.

Baird, D.T. (1977b). 'Synthesis and secretion of steroid hormones by the ovary *in vivo*.' In *The Ovary*, Vol. III, 2nd edition (eds. S. Zuckerman and B.J. Weir), pp. 305–357. Academic Press, London.

Baird, D.T., and Fraser, I.S. (1974). 'Blood production and ovarian secretion rates of oestradiol-17β and oestrone in women throughout the menstrual cycle.' *J. Clin. Endocrinol. Metab.*, **38**, 1009–1017.

Baird, D.T., Baker, T.G., McNatty, K.P., and Neal, P. (1975). 'Relationship between the secretion of the corpus luteum and the length of the follicular phase of the ovarian cycle.' *J. Reprod. Fertil.*, **45**, 611–619.

Birnbaumer, L., and Kirchik, H.J. (1983). 'Regulation of gonadotrophic action: the molecular mechanisms of gonadotrophin induced activation of ovarian adenylyl cyclases.' In *Factors Regulating Ovarian Function* (eds. G.S. Greenwald and P.F. Terranova), pp. 49–54. Raven Press, New York.

Bjersing, L. (1981). 'Correlation of fine structure and endocrine function in the human corpus luteum.' In *Functional Morphology of the Human Ovary* (ed. J.R.T. Coutts), pp. 119–139. MTP Press, Lancaster.

Carr, B.R., Sadler, R.K., Rochelle, D.B., Stalmach, M.A., MacDonald, P.C., and Simpson, E.R. (1981). 'Plasma lipoprotein regulation of progesterone biosynthesis by human corpus luteum tissue in organ culture.' *J. Clin. Endocrinol. Metab.*, **52**, 875–881.

Carr, B.R., MacDonald, P.C., and Simpson, E.R. (1982). 'The role of lipoproteins in the regulation of progesterone secretion by the human corpus luteum.' *Fertil. Steril.*, **38**, 303–311.

Channing, C.P. (1980). 'Progesterone and estrogen secretion by cultured monkey ovarian cell types: influences of follicular size, serum luteinizing hormone levels and follicular fluid estrogen levels.' *Endocrinology*, **107**, 342–352.

Channing, C.P., and Reichert, L.E., Jr. (1983). 'Effect of exposure of the monkey follicle to the midcycle LH/FSH surge upon the ability of granulosa and theca to respond to FSH *in vitro* with increased estrogen and progesterone secretion.' In *Factors Regulating Ovarian Function* (eds. G.S. Greenwald and P.F. Terranova), pp. 329–332. Raven Press, New York.

Channing, C.P., Thanki, K., Lindsey, A.M., and Ledwitz-Rigby, F. (1978). 'Development and hormonal regulation of gonadotrophin responsiveness in granulosa cells of the mammalian ovary.' In *Receptors and Hormone Action* (eds. L. Birnbaumer and B.W. O'Malley), pp. 435–455. Academic Press, New York.

Dennefors, B.L., Hamberger, L., and Nilsson, L. (1983). 'Influence of human chorionic gonadotrophin *in vivo* on steroid formation and gonadotrophin responsiveness in isolated human preovulatory follicle cells.' *Fertil. Steril.*, **39**, 56–61.

di Zerega, G.S., and Hodgen, G.D. (1980). 'Changing functional status of the monkey corpus luteum.' *Biol. Reprod.*, **23**, 253–263.

Djahanbakhch, O., McNeilly, A.S., Warner, P.M., Swanston, I.A., and Baird, D.T. (1983). 'Changes in plasma levels of prolactin, FSH, oestradiol, androstenedione and progesterone around the preovulatory surge of LH in women.' *Clin. Endocrinol.*, **20**, 463–472.

Dorrington, J.H., McKercher, H.L., Chan, A.K., and Gore-Langton, R.E. (1983). 'Hormonal interactions in the control of granulosa cell differentiation.' *J. Steroid Biochem.*, **19**, 17–32.

Edwards, R.G. (1974). 'Follicular fluid.' *J. Reprod. Fertil.*, **37**, 189–219.

Edwards, R.G., Steptoe, P.C., Fowler, R.E., and Baillie, J. (1980). 'Observations on preovulatory human ovarian follicles and their aspirates.' *Br. J. Obstet. Gynaecol.*, **87**, 769–779.

Fritz, M.A., and Speroff, L. (1982). 'The endocrinology of the menstrual cycle: the interaction of folliculogenesis and neuroendocrine mechanisms.' *Fertil. Steril.*, **38**, 509–529.

Fujita, Y., Mori, T., Suzuki, A., Nihnobu, K., and Nishimura, T. (1981). 'Functional and structural relationships in steroidogenesis *in vitro* by human corpora lutea during development and regression.' *J. Clin. Endocrinol. Metab.*, **53**, 744–751.

Goldstein, J.L., and Brown, M.S. (1977). 'The low-density lipoprotein pathway and its relation to artheroschlerosis.' *Annu. Rev. Biochem.*, **46**, 897–930.

Gospodarowicz, D., and Thakral, K.K. (1978). 'Production of corpus luteum angiogenic factor responsible for proliferation of capillaries and neovascularization of the corpus luteum.' *Proc. Natl. Acad. Sci. USA*, **75**, 847–851.

Gougeon, A., and Lefevre, B. (1983). 'Evolution of the diameters of the largest healthy and atretic follicles during the human menstrual cycle.' *J. Reprod. Fertil.*, **69**, 497–502.

Gwynne, J.T., and Strauss, J.F. (1982). 'The role of lipoproteins in steroidogenesis and cholesterol metabolism in steroidogenic glands.' *Endocrinol. Rev.*, **3**, 299–329.

Harrison, R.J., and Weir, B.J. (1977). 'Structure of the mammalian ovary.' In *The Ovary*, Vol. I, 2nd edition (eds. S. Zuckerman and B.J. Weir), pp. 113–217. Academic Press, London.

Hillier, S.G. (1981). 'Regulation of follicular oestrogen biosynthesis: a survey of current concepts.' *J. Endocrinol.*, **89**, 3P–18P.

Hillier, S.G., Zeleznik, A.J., and Ross, G.T. (1978). 'Independence of steroidogenic capacity and luteinizing hormone receptor induction in developing granulosa cells.' *Endocrinology*, **102**, 937–946.

Hillier, S.G., Reichert, L.E., Jr., and Van Hall, E.V. (1981). 'Control of preovulatory follicular estrogen biosynthesis in the human ovary.' *J. Clin. Endocrinol. Metab.*, **52**, 847–856.

Hillier, S.G., Trimbos-Kemper, T.C.M., Reichert, L.E., Jr., and Van Hall, E.V. (1983). 'Gonadotrophic control of human granulosa cell function: *in-vitro* studies on aspirates of preovulatory follicles.' In *Factors Regulating Ovarian Function* (eds. G.S. Greenwald and P.F. Terranova), pp. 49–54. Raven Press, New York.

Hillier, S.G., Parsons, J.H., Margara, R.A., Winston, R.M.L., and Crofton, M.E. (1984). 'Serum oestradiol and preovulatory follicular development before *in-vitro* fertilization.' *J. Endocrinol.*, **101**, 113–118.

Hoff, J.D., Quigley, M.E., and Yen, S.S.C. (1983). 'Hormonal dynamics at midcycle: a reevaluation.' *J. Clin. Endocrinol. Metab.*, **57**, 792–796.

Hsueh, A.J.W., Jones, P.B.C., Adashi, E.Y., Wang, C., Zhuang, L.-Z., and Welsh, T.H., Jr. (1983). 'Intraovarian mechanisms in the hormonal control of granulosa cell differentiation in rats.' *J. Reprod. Fertil.*, **69**, 325–342.

Hunter, M.G., and Baker, T.G. (1981). 'Effect of hCG, cAMP and FSH on steroidogenesis by human corpora lutea *in vitro*.' *J. Reprod. Fertil.*, **63**, 285–288.

Hutchison, J.S., and Zeleznik, A.J. (1983). 'The rhesus monkey corpus luteum is dependent upon pituitary gonadotrophin secretion throughout the luteal phase of the menstrual cycle.' *Biol. Reprod. Suppl.*, **1**, abstract 94.

Kerin, J., Edmonds, D.K., Warnes, C.M., Cox, L.W., Seamark, R.F., Matthews, C.D., Young, G.B., and Baird, D.T. (1981). 'Morphological and functional relations of Graafian follicle growth to ovulation in women using ultrasonic, laproscopic and biochemical measurements.' *Br. J. Obstet. Gynaecol.*, **88**, 81–90.

Leung, P.C.K., and Armstrong, D.T. (1980). 'Interactions of steroids and gonadotrophins in the control of steroidogenesis in the ovarian follicle.' *Annu. Rev. Physiol.*, **42**, 71–82.

Macnaughton, M.C., Samad Kader, A., Gaukroger, J.M., and Coutts, J.R.T. (1981). 'Steroid hormone production by the human corpus luteum.' In *Functional Morphology of the Human Ovary* (ed. J.R.T. Coutts), pp. 137–156. MTP Press, Lancaster.

Marsh, J.M. (1976). 'The role of cyclic AMP in gonadal steroidogenesis.' *Biol. Reprod.*, **14**, 30–53.

McNatty, K.P. (1978). 'Follicular fluid.' In *The Vertebrate Ovary* (ed. R.E. Jones), pp. 215–259. Plenum Press, New York.

McNatty, K.P. (1982). 'Ovarian follicular development from the onset of luteal regression in humans and sheep.' In *Follicular Maturation and Ovulation* (*Excerpta Medica International Congress Series* 560) (eds. R. Rolland, E.V. Van Hall, S.G. Hillier, K.P. McNatty and J. Schoemaker), pp. 1–18. Excerpta Medica, Amsterdam.

McNatty, K.P., Makris, A., De Grazia, C., Osathanondh, R., and Ryan, K.J. (1979). 'The production of progesterone, androgens and estrogens by granulosa cells, thecal tissue, and stromal tissue from human ovaries *in vitro*.' *J. Clin. Endocrinol. Metab.*, **49**, 687–699.

McNatty, K.P., Makris, A., Osathanondh, R., and Ryan, K.J. (1980). 'Effects of luteinizing hormone on steroidogenesis by thecal tissue from human ovarian follicles *in vitro*.' *Steroids*, **36**, 53–63.

McNatty, K.P., Hillier, S.G., Van Den Boogaard, A.M.J., Trimbos-Kemper, T.C.M., Reichert, L.E., Jr., and Van Hall, E.V. (1983). 'Follicular development during the luteal phase of the human menstrual cycle.' *J. Clin. Endocrinol. Metab.*, **56**, 1022–1031.

McNeilly, A.S., Kerin, J., Swanston, I.A., Bramley, T.A., and Baird, D.T. (1980). 'Changes in the binding of human chorionic gonadotrophin, follicle-stimulating hormone and prolactin to human corpora lutea during the menstrual cycle and pregnancy.' *J. Endocrinol.*, **87**, 315–325.

Moor, R.M. (1974). 'The ovarian follicle of the sheep: inhibition of oestrogen secretion by luteinizing hormone.' *J. Endocrinol.*, **61**, 455–463.

Moor, R.M. (1977). 'Sites of steroid production in ovine Graafian follicles.' *J. Endocrinol.*, **73**, 143–150.

Mori, T., Nihnobu, K., Takeuchi, S., Ohno, Y., and Tojo, S. (1983). 'Interrelation between luteal cell types in steroidogenesis *in vitro* of human corpus luteum.' *J. Steroid Biochem.*, **19**, 811–815.

Ohashi, M., Carr, B.R., and Simpson, E.R. (1982). 'Lipoprotein-binding sites in human corpus luteum membrane fractions.' *Endocrinology*, **110**, 1477–1482.

Reichert, L.E., Jr., (1967). 'Selective inactivation of the luteinizing hormone containment of follicle-stimulating hormone preparations by digestion with α-chymotrypsin.' *J. Clin. Endocrinol. Metab.*, **27**, 1065–1067.

Reynolds, S.R.M. (1973). 'Blood and lymph vascular systems of the ovary.' In *Handbook of Physiology*, Section 7, Vol. II (ed. R.O. Greep), pp. 261–316. American Physiological Society, Washington, DC.

Richards, J.S. (1980). 'Maturation of ovarian follicles: actions and interactions of pituitary and ovarian hormones on follicular differentiation.' *Physiol. Rev.*, **60**, 51–89.

Ross, G.T., and Hillier, S.G. (1978). 'Luteal maturation and luteal phase defect.' *Clin. Obstet. Gynecol.*, **5**, 391–409.

Ryan, K.J. (1979). 'Granulosa–thecal cell interactions in ovarian steroidogenesis.' *J. Steroid Biochem.*, **11**, 799–800.

Shalgi, R., Kraicer, P., Rimon, A., Pinto, M., and Soferman, N. (1973). 'Proteins of human follicular fluid: the blood–follicle barrier.' *Fertil. Steril.*, **24**, 429–434.

Simpson, E.R., Rochelle, D.B., Carr, B.R., and MacDonald, P.C. (1980). 'Plasma lipoproteins in follicular fluid of human ovaries.' *J. Clin. Endocrinol. Metab.*, **51**, 1469–1471.

Testart, J., Castanier, M., Feinstein, M.-C., and Frydman, R. (1982a). 'Pituitary and steroid hormones in the preovulatory follicle during spontaneous or stimulated cycles.' In *Follicular Maturation and Ovulation* (*Excerpta Medica Congress Series* 560) (eds. R. Rolland, E.V. Van Hall, S.G. Hillier, K.P. McNatty and J. Schoemaker), pp. 193–201. Excerpta Medica, Amsterdam.

Testart, J., Frydman, R., Nahoul, K., Grenier, J., Feinstein, M.-C., Roger, M., and Scholler, R. (1982b). 'Steroids and gonadotrophins during the late preovulatory phase of the menstrual cycle. Time relations between plasma hormone levels and luteinizing hormone surge onset.' *J. Steroid Biochem.*, **17**, 675–682.

Tsang, B.K., Moon, Y.S., Simpson, C.W., and Armstrong, D.T. (1979). 'Androgen biosynthesis in human ovarian follicles: cellular source, gonadotrophic control, and adenosine-3′, 5′-monophosphate mediation.' *J. Clin. Endocrinol. Metab.*, **48**, 153–158.

Tureck, R.W., and Strauss, J.F. (1982). 'Progesterone synthesis by luteinized human granulosa cells in culture: The role of *de novo* sterol synthesis and lipoprotein-carried sterol.' *J. Clin. Endocrinol. Metab.*, **54**, 367–373.

Van Look, P.F.A., and Baird, D.T. (1980). 'Regulatory mechanisms during the menstrual cycle.' *Eur. J. Obstet. Gynecol. Reprod. Biol.*, **11**, 121–144.

World Health Organization (1981). 'Temporal relationships between ovulation and defined changes in the concentration of plasma estradiol-17β, luteinizing hormone, follicle-stimulating hormone, and progesterone. I. Probit analysis.' *Am. J. Obstet. Gynecol.*, **138**, 383–390.

World Health Organization (1981). 'Temporal relationships between ovulation and defined changes in the concentration of plasma estradiol-17β, luteinizing hormone, follicle-stimulating hormone, and progesterone. II. Histological dating.' *Am. J. Obstet. Gynecol.*, **139**, 886–895.

Young, W.C. (1961). 'The mammalian ovary.' In *Sex and Internal Secretions*, Vol. I, 3rd edition (ed. W.C. Young), pp. 449–496. Ballière, Tindall & Cox, London.

Zeleznik, A.J., Midgley, A.R., Jr., and Reichert, L.E., Jr. (1974). 'Granulosa cell maturation in the rat: increased binding of human chorionic gonadotrophin following treatment with follicle-stimulating hormone *in vivo*.' *Endocrinology*, **95**, 818–825.

The Luteal Phase
Edited by S.L. Jeffcoate
© 1985 John Wiley & Sons Ltd.

CHAPTER 2

Control of luteolysis

DAVID T. BAIRD
Department of Obstetrics and Gynaecology,
University of Edinburgh,
Centre for Reproductive Biology,
37 Chalmers Street,
Edinburgh EH3 9EW, UK

Introduction

The corpus luteum is formed in response to the mid-cycle surge of luteinizing hormone (LH) from the remnants of the follicle following ovulation. Progesterone, its principal secretory product, provides the hormonal milieu necessary for the continued development of the fertilized ovum through its action on the uterus to maintain an environment which is favourable for implantation of the blastocyst about 6 or 7 days after fertilization. The trophoblast of the successfully implanted blastocyst secretes chorionic gonadotrophin (hCG) which maintains the structure and secretory capacity of the corpus luteum beyond the time of the missed menstrual period. It is clear therefore that the correct function of the corpus luteum is essential if pregnancy is going to continue (Rothchild, 1981).

In spite of its key role in the process of reproduction we still do not fully understand the functioning of the human corpus luteum. In particular the important process (luteolysis) by which the corpus luteum ceases to function 10–12 days after its formation in the absence of pregnancy is a complete mystery in the human. Is the corpus luteum starved of luteotrophic support at the time of luteal regression or is it destroyed by the production of a luteolytic agent as in many animal species? In this chapter, I shall summarize our current knowledge of the control of luteolysis in the human although it will be necessary to refer to studies in other primates and the mechanism in animal species in which prostaglandin $F_{2\alpha}$ ($PGF_{2\alpha}$) has been shown to be the luteolytic agent.

First, it is necessary to consider the factors responsible for the formation and maintenance of the corpus luteum.

1. Formation and Maintenance of the Corpus Luteum

The preovulatory surge of LH acts on the granulosa cells of the mature Graafian follicle to induce a series of morphological and biochemical changes known as luteinization (see Chapter 1). Prior to exposure to the LH surge steroidogenic activity has been almost exclusively confined to aromatization of androgens to oestrogens.

The hormonal requirements for luteinization have been studied by culturing granulosa cells *in vitro*. LH and probably prolactin are required for maximum steroidogenesis (McNatty *et al.*, 1974). LH interacts with specific receptors on the cell membrane to change the granulosa into luteal cells which secrete progesterone (Gibori and Miller, 1982). It does this by activation of the enzyme

STEROID SYNTHESIS IN LUTEAL CELL

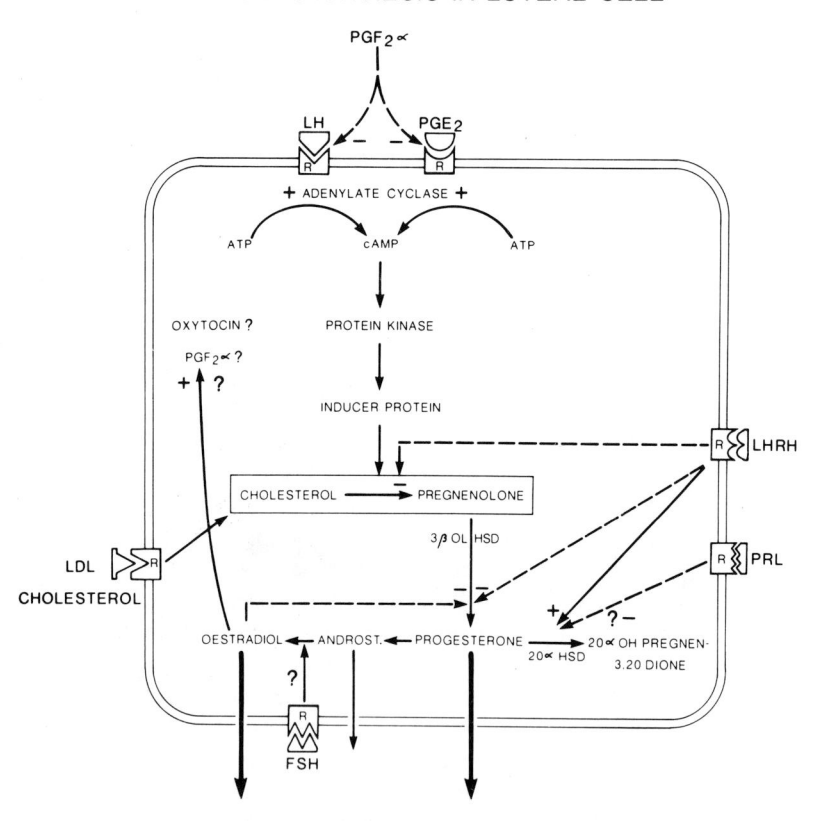

Figure 1. Factors regulating the synthesis of steroids by the luteal cell. LH interacts with its receptor (R) to stimulate the production of cAMP and eventually the conversion of cholesterol to pregnenolone within the mitochondrion. Although receptors (R) for LHRH, FSH and PRL (prolactin) have been identified in the human luteal cell their physiological role is still unknown. (HSD = hydroxysteroid dehydrogenase)

adenylate cyclase which converts ATP to cyclic AMP (cAMP) which is the 'second messenger' in transducing the trophic effect (Marsh, 1976) (Figure 1; and cf. Figure 15 in Chapter 1).

1.1. Morphology of the Corpus Luteum

The human corpus luteum is composed of at least two cell types (Figure 2; see also Figures 7 and 10 in Chapter 1). Theca-lutein cells together with invading blood vessels divide the granulosa-lutein cells into a series of islands. Both granulosa-lutein and theca-lutein cells contain abundant tubular smooth endoplasmic reticulum, mitochondria with tubular cristae, lipid droplets and numerous free ribosomes (Corner, 1956; Gillim et al., 1969). As these ultrastructural features are characteristic of steroid-secreting cells it is assumed that both cell types are involved in steroid synthesis, although studies on the biosynthetic activity of individual cell types have not been performed. In addition, prominent macrophages (K cells) surround the capillaries.

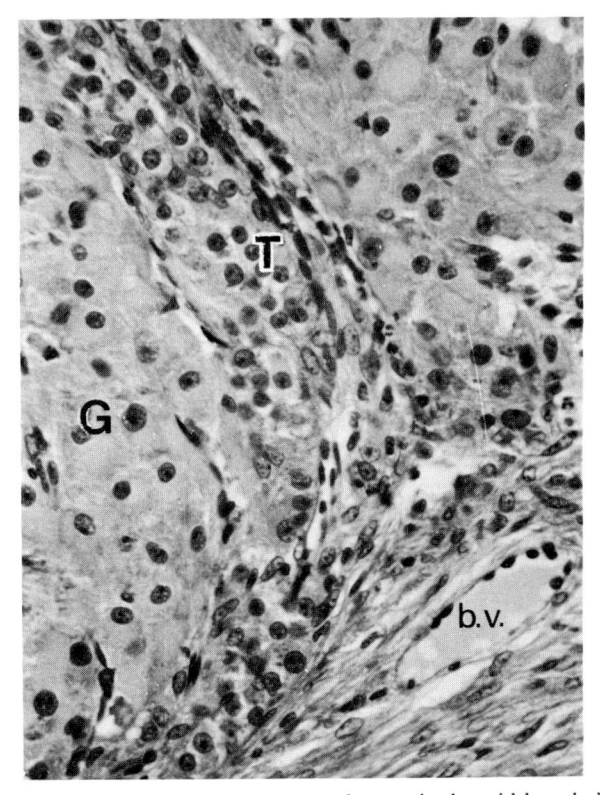

Figure 2. Photomicrograph of a human corpus luteum in the mid-luteal phase of the cycle showing the distinct granulosa (G) and theca (T) luteal cells. b.v. = blood vessel

1.2. Secretions of the Corpus Luteum

The primate corpus luteum, including that of man, is unusual in the variety of steroid hormones secreted. In addition to progesterone, the following are all synthesized hy human luteal tissue and secreted in the ovarian vein: pregnenolone, 17β-hydroxyprogesterone, androstenedione, testosterone, dehydroepiandrosterone, oestrone and oestradiol. In women the corpus luteum secretes almost as much oestrogen as the preovulatory follicle and is virtually the sole source of oestradiol and progesterone in the second half of the cycle.

There is now convincing evidence that the quantity of progesterone secreted *in vivo* is determined not only by LH but by the amount of cholesterol available as substrate (Carr *et al.*, 1981) (see section 3.1 in Chapter 1). Only that fraction of cholesterol in plasma which is bound to low density lipoprotein (LDL) can enter the luteal cell in any significant quantities via a specific LDL receptor mechanism (Figure 1). The conversion of cholesterol to pregnenolone within the mitochondrion is stimulated by LH with cytochrome P-450 playing a key role in the side-chain cleavage of cholesterol (Marsh, 1976). Pregnenolone is then converted to progesterone by the enzyme 3β-ol steroid dehydrogenase within the cytoplasm before being secreted into the extracellular fluid and the blood stream. Progesterone probably leaves the cells by diffusing down a concentration gradient, although it has been suggested that it is packaged in granules which are secreted from the cell surface by exocytosis (Gemmell *et al.*, 1974).

1.3. Luteotrophic Support of the Corpus Luteum

In most mammalian species studied LH is necessary for steroid secretion by the corpus luteum. In women LH receptors of high affinity and low capacity have been identified and characterized at various stages of the cycle (McNeilly *et al.*, 1980; Rajaniemi *et al.*, 1981). However, the evidence that endogenous pituitary LH is required for maintenance of the corpus luteum in women is scanty and contradictory. When ovulation is stimulated by gonadotrophins in hypophysectomized women, normal luteal function results after only a single ovulation dose of hCG (Van de Wiele *et al.*, 1970), suggesting that once formed the human corpus luteum requires no further luteotrophic support from the anterior pituitary. However, the half-life of hCG in plasma and probably on the LH receptor is much longer than that of pituitary LH. In a patient in whom ovulation was induced with an infusion of LH to mimic the endogenous surge of pituitary LH, the secretion of progesterone was reduced and the luteal phase only lasted 5 days (Figure 3). These studies suggest that in women the corpus luteum is relatively independent of pituitary luteotrophic support and requires LH only for a few days in the early luteal phase of the cycle as in the sheep.

There are no data on the acute effect of hypophysectomy on the functioning of the corpus luteum in women. Experimental studies in the monkey are confusing. In rhesus monkeys hypophysectomy soon after ovulation resulted in a

CONTROL OF LUTEOLYSIS

INDUTION OF OVULATION IN HYPOPHYSECTOMIZED WOMAN

LH 1·2 U/l   FSH 1·2 U/l   PR 160 mU/l

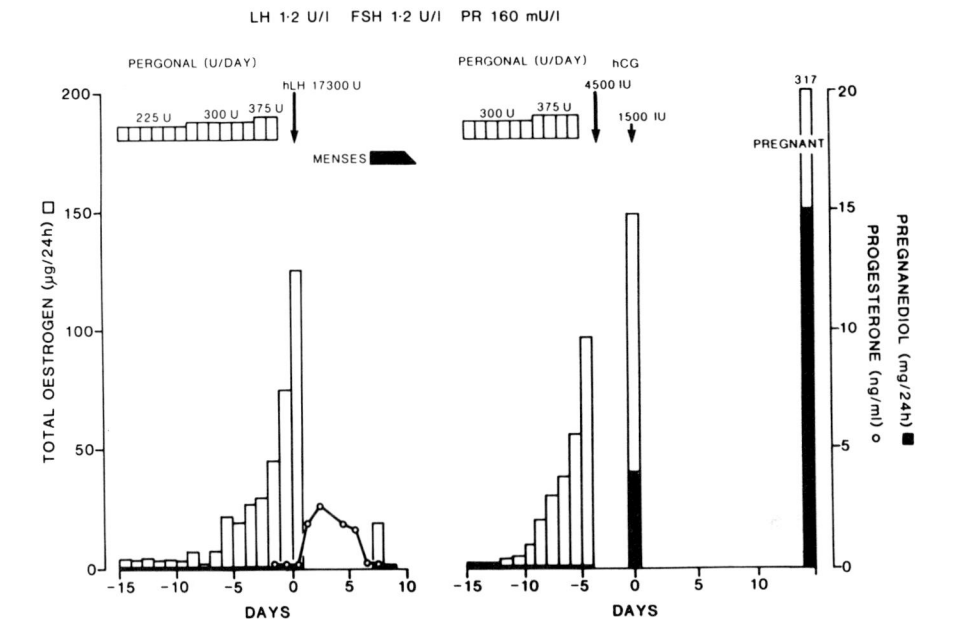

Figure 3. Induction of ovulation by gonadotrophins in a hypogonadotrophic woman. Following total hypophysectomy for acromegaly 3 years previously the patient had remained amenorrhoeic due to very low levels of gonadotrophins (LH = 1.2 U/l, FSH = 1.2 U/l). Follicular development as indicated by the increased excretion of total oestrogen in the urine was induced by daily injection of human menopausal gonadotrophin (Pergonal). In the first cycle when ovulation was induced with an infusion of LH (17 300 U) over 36 h, the progesterone concentration never exceeded 4 ng/ml and menstruation started 6 days later. In the second cycle when hCG 4500 IU was used to induce ovulation a normal luteal phase occurred and the woman became pregnant. (D.T. Baird and A.S. McNeilly, unpublished data)

luteal phase of nearly normal length (Asch et al., 1982) (Figure 4). However, experiments in which endogenous LH was neutralized by the injection of an antiserum to hCG which cross-reacted with monkey LH, resulted in suppression of progesterone concentration (Mougdal et al., 1971, 1972). These studies suggest that at least up to the mid-luteal phase the corpus luteum requires the presence of LH continuously. The reason for this discrepancy is not clear and further studies are needed of the luteotrophic requirements of the corpus luteum at different stages of the luteal phase.

During the luteal phase of the cycle there is a progressive decline in the concentration of LH in plasma (Midgley and Jaffe, 1968). The frequency of LH

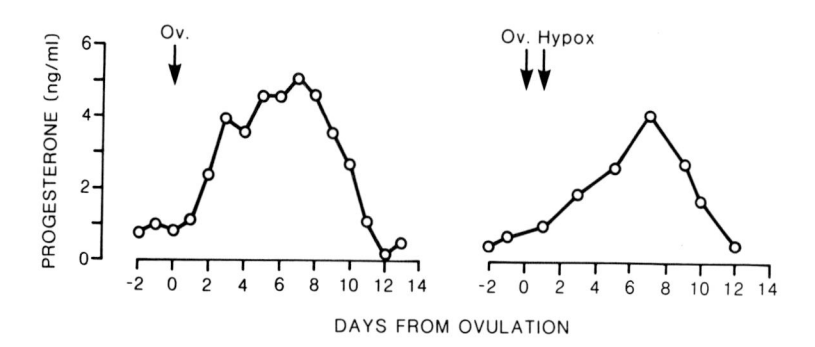

Figure 4. Concentration of progesterone in rhesus monkey hypophysectomized (Hypox) on the day after ovulation (Ov.). Note that although the levels of progesterone are lower after hypophysectomy the length of the luteal phase is similar to that in the control cycle in the left side of the figure. (Redrawn from Asch et al., 1982)

pulses slows to approximately one pulse every 3 h, although the relative amplitude of each pulse is increased (Santen and Bardin, 1973; Backstrom et al., 1982). Following each pulse of LH there is a corresponding increase in the concentration of progesterone and oestradiol, supporting the view that LH provides continuing luteotrophic support to the corpus luteum.

The corpus luteum also contains receptors for prolactin and follicle-stimulating hormone (FSH) (Poindexter et al., 1979; McNeilly et al., 1980). The role of prolactin in maintenance of the corpus luteum is discussed in detail in Chapter 4. The type of luteal cell which contains FSH receptors is unknown as is its function. The synthesis of oestradiol but not progesterone by minces of luteal tissue is stimulated by the addition of FSH, suggesting that as in the follicle aromatization may be influenced by this hormone (Hunter, 1980).

2. Luteal Regression by Withdrawal of Luteotrophic Support

It seems likely that the corpus luteum in the human is dependent at least up to the mid-luteal phase on the secretions of the anterior pituitary. In the baboon and marmoset monkey, injection of antiserum to hCG in early pregnancy will cause luteal regression (Hearn, 1980) and, in women, the life span of the corpus luteum in the non-pregnant cycle can be extended by injection of exogenous hCG (Hanson et al., 1971). The fact that in early pregnancy the corpus luteum is rescued by the secretion of hCG from the trophoblast might suggest that there is a relative lack of LH in the late luteal phase of the normal cycle. Although the concentration of LH is suppressed to low levels in the mid-luteal phase of the cycle, it seems unlikely that luteal regression is due solely to suppression of pituitary LH. In a negative feedback system any reduction in LH secretion

hCG/LH BINDING IN HUMAN CORPUS LUTEUM

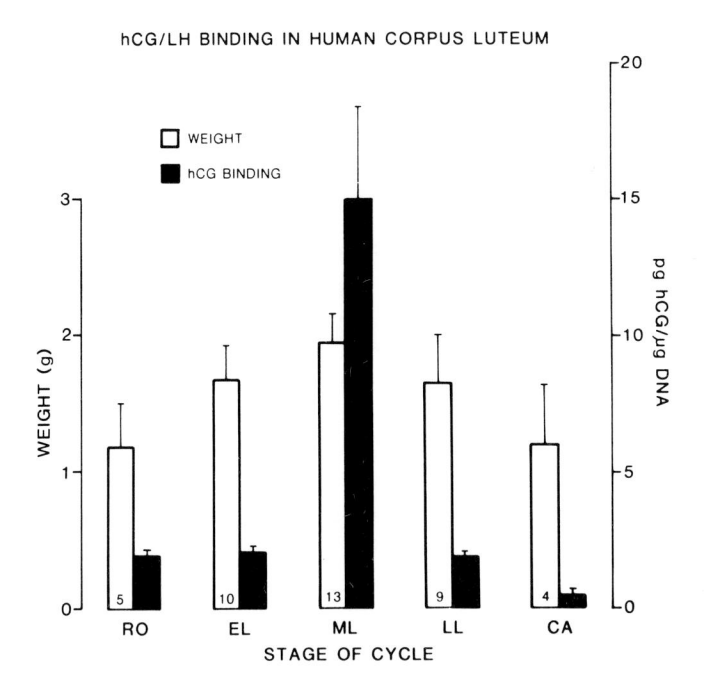

Figure 5. Binding of hCG to human corpus luteum at different stages of the luteal phase. Each bar represents the mean (±SE) of the number of samples as indicated in the column. RO = recent ovulation; EL, ML and LL = early, mid- and late luteal; CA = corpus albicans. (From McNeilly et al., 1980)

should result in a reduced secretion of oestradiol and progesterone by the corpus luteum and a consequent increased secretion of LH (Baird et al., 1975). Thus the low levels of LH probably contribute to a luteolysis by making the corpus luteum susceptible to luteolytic agents.

The binding of hCG/LH to the corpus luteum varies depending on its age (Figure 5). The low binding to the early corpus luteum probably indicates desensitization and/or occupancy of the receptors following the mid-cycle LH surge. However, the marked decline in receptors in the late luteal phase may be responsible in fact for the decline in progesterone secretion at this time.

3. Production of a Luteolytic Agent

3.1. Prostaglandin $F_{2\alpha}$

In many animals (e.g. sheep, guinea pig, cow) luteolysis is induced by the secretion of prostaglandin $F_{2\alpha}$ ($PGF_{2\alpha}$) from the uterus (Horton and Poyser, 1976: Baird, 1978). It has been known for many years that in these species hysterectomy results in prolongation of the life span of the corpus luteum. Over 10 years ago it was discovered that $PGF_{2\alpha}$ was released from the uterus at the end of the cycle and reached the adjacent ovary by counter-current transfer from the uterine vein to the ovarian artery (McCracken et al., 1971).

$PGF_{2\alpha}$ is thought to induce regression of the corpus luteum by interfering with the ability of LH to activate adenylate cyclase (Henderson and McNatty, 1975; Grinwich et al., 1976; Hamberger et al., 1979). The corpus luteum is only susceptible to the luteolytic action of $PGF_{2\alpha}$ some days after its formation (Baird, 1978). In the pig, for example, the corpus luteum is resistant to $PGF_{2\alpha}$ until day 14 while $PGF_{2\alpha}$ is luteolytic in the sheep and cow after day 4. This period of resistance to $PGF_{2\alpha}$ corresponds to the time when hypophysectomy has little effect on progesterone secretion and suggests that there is probably a period of time, characteristic of each species, when the function of the corpus luteum is relatively autonomous. It was originally suggested by Henderson and McNatty (1975) that $PGF_{2\alpha}$ gained access to its receptor only when the levels of LH in blood were relatively low and a significant number of the LH receptors were unoccupied. Thus the corpus luteum would only be susceptible to $PGF_{2\alpha}$ in the mid- and late luteal phase of the cycle.

The luteolytic properties of $PGF_{2\alpha}$ and its analogues have been used to synchronize oestrous cycles in sheep, cows and horses (Cooper and Walpole, 1975). For this reason extensive research has been carried out in women to determine whether $PGF_{2\alpha}$ is luteolytic and if so whether it is involved in the normal mechanism of luteal regression.

In women and monkeys hysterectomy has no effect on cyclicity so that if $PGF_{2\alpha}$ is involved in luteal regression it cannot be derived from the uterus (Beling et al., 1970; Knobil, 1973). Studies designed to determine whether $PGF_{2\alpha}$ or its potent analogues are luteolytic *in vivo* in primates are confusing and often conflicting (Kirton, 1975). The confusion has probably arisen because of the difficulty (associated with unacceptable side-effects) of delivering adequate quantities of $PGF_{2\alpha}$ to the ovary by systemic infusion. In addition, in many studies insufficient care was taken to control the stage of the luteal phase when the experiments were conducted. As mentioned earlier, it is likely that in the primate, as in other species, the corpus luteum will be resistant to the luteolytic effect of $PGF_{2\alpha}$ for at least some time after formation. Since it is likely to be more susceptible to $PGF_{2\alpha}$ as it ages it becomes difficult to demonstrate a significant shortening of the luteal phase.

However, several studies have shown either a temporary fall in the concentration of progesterone during and after the infusion of $PGF_{2\alpha}$ (Wentz and Jones, 1973) or after the administration by vaginal pessary of a potent analogue (Hamberger et al., 1980). In another study in which 500–1000 µg of $PGF_{2\alpha}$ were injected directly into the corpus luteum at laparoscopy there was a shortening of the luteal phase (Korda et al., 1975). These studies need repeating with large numbers at specific stages of the luteal phase. Evidence from luteinized granulosa cells or dispersed luteal cells *in vitro* add further support to the concept that $PGF_{2\alpha}$ is luteolytic in man (McNatty *et al.*, 1975). In a series of careful experiments it was demonstrated that $PGF_{2\alpha}$ inhibited the ability of LH or hCG to stimulate the production of progesterone and cAMP from dispersed luteal cells obtained from corpora lutea in the mid-luteal phase. In contrast, cells from the early luteal phase were resistant to $PGF_{2\alpha}$ (Dennefors *et al.*, 1982).

These studies demonstrate that $PGF_{2\alpha}$ can act on the human luteal cell to suppress synthesis of progesterone in a manner similar to its action in those species like the rat where it is the luteolytic agent. Receptors for $PGF_{2\alpha}$ are present on human luteal cells (Powell *et al.*, 1974). Furthermore, several studies have demonstrated that luteal tissue can synthesize prostaglandins and the concentration of $PGF_{2\alpha}$ in the human corpus luteum is relatively high (Swanston *et al.*, 1977). However, although one study claimed that the concentration of $PGF_{2\alpha}$ rose in the late luteal phase (Shutt *et al.*, 1976) we were unable to demonstrate any significant rise before the onset of luteal regression (Figure 6) (Swanston *et al.*, 1977; see also Challis *et al.*, 1976; Patwardhan and Lanthier, 1980). These studies, however, do not necessarily exclude a role for prostaglandins in luteal regression in the primate because the corpus luteum becomes increasingly vulnerable to the luteolytic effect of $PGF_{2\alpha}$ as it ages. Moreover, PGE_2 is known to stimulate progesterone secretion *in vitro* and in the sheep can prevent the luteolytic effect of $PGF_{2\alpha}$ *in vivo* (Henderson *et al.*, 1977). Studies in the rhesus monkey have demonstrated a significant fall in the production of PGE in the mid-luteal phase of the cycle so that the ratio of $PGF_{2\alpha}$ to PGE exceeds unity

CONCENTRATION OF PGF2∝ IN HUMAN CORPUS LUTEUM

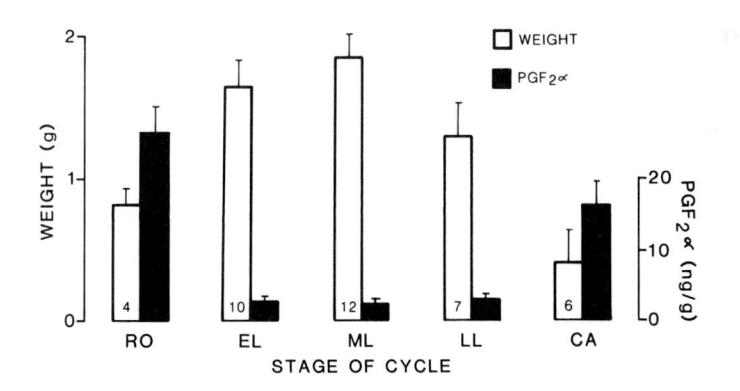

Figure 6. Concentration of $PGF_{2\alpha}$ in human corpus luteum at different stges of the luteal phase. RO = recent ovulation; EL, ML and LL = early, mid- and late luteal phase; CA = corpus albicans. Each bar is the mean (± SE) of the number indicated in the column. (From Swanston *et al.*, 1977)

Table 1. Prostaglandin production by corpus luteum of rhesus monkey

Luteal phase	Ratio of PGF/PGE
Early ($n = 5$)	0.2 ± 0.05
Mid ($n = 5$)	0.65 ± 0.05
Late ($n = 5$)	3.78 ± 1.17

Data from Balmaceda *et al.* (1979).

(Balmeceda *et al.*, 1979) (Table 1). However, the fact that the life span of the corpus luteum is apparently unaffected when prostaglandin production is inhibited by administration of indomethacin, leaves doubt as to the physiological role of prostaglandins in luteal regression (Manaugh and Novy, 1976).

3.2. Oestrogen

It is well established that in women administration of oestrogen in the luteal phase of the cycle causes suppression of progesterone secretion and shortens the luteal phase (Johansson and Gemzell, 1971). Injection of oestradiol but not testosterone directly into the ovary containing the corpus luteum provokes premature onset of menstrual bleeding (Hoffmann, 1960). It is likely that oestradiol in combination with progesterone exerts a strong suppression of the secretion of LH (and FSH) from the pituitary (Baird *et al.*, 1975). However, the fact that local injection of oestradiol is only effective when injected into the ovary containing the corpus luteum, raises the possibility that oestradiol may be acting within the corpus luteum itself (Karsch and Sutton, 1976).

The maximum secretion of both progesterone and oestradiol by the corpus luteum is achieved about 7–10 days after formation. The concentraion of progesterone in luteal tissue is lower in the mid-luteal than in the early luteal phase, presumably due to the more rapid removal of progesterone associated with a better developed blood supply (Swanston *et al.*, 1977) (Figure 7). In contrast, the concentration of oestradiol rises in the mid-luteal phase so that the ratio of progesterone to oestradiol falls from about 200/1 to 55/1. The reason for this change in the relative concentration of steroids is unknown but it may be due to the presence of a specific binding protein for oestradiol such as is present in the rabbit (Keyes *et al.*, 1983). There is a similar change in the ratio of androstenedione to oestradiol suggesting that aromatization of androgens is more efficient. Whatever the reason the biological consequence is a marked change in the hormonal environment within the corpus luteum similar to that which favours the release of prostaglandins from the endometrium (Baird, 1978).

Recent studies in rhesus monkeys have re-investigated the mechanism of

CONCENTRATION OF PROGESTERONE AND OESTRADIOL IN HUMAN CORPUS LUTEUM

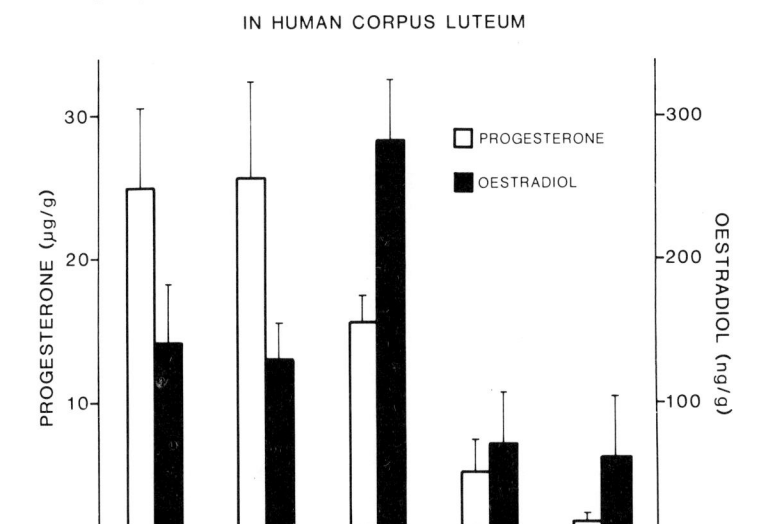

Figure 7. Concentration of progesterone and oestradiol in the human corpus luteum. See legend to Figure 5 for details. (From Swanston et al., 1977)

luteolytic effect of oestradiol (Schoonmaker et al., 1981, 1982). Oestrogen induces premature regression of the corpus luteum in the rhesus monkey when administered systemically or by injection into the corpus luteum only when given during the mid- to late luteal phase but is ineffective when given 2–6 days after the LH surge. It would appear, therefore, that the primate corpus luteum is initially refractory to the luteolytic action of oestradiol (just as we have seen for $PGF_{2\alpha}$) and the ability to respond to oestradiol is acquired as the luteal phase progresses.

What is the mechanism for oestradiol-induced luteal regression? When oestradiol is implanted locally within the corpus luteum premature regression of the corpus luteum is associated with a significant depression of LH concentration (Schoonmaker et al., 1982) and luteolysis only occurs in those monkeys in which the secretion of LH falls. Thus it seems likely that if oestradiol has a local action within the corpus luteum it also requires a systemic action on pituitary LH. In this context it is unlikely that elevation of LH alone will prolong luteolysis because administration of an anti-oestrogen had no effect on the onset of menstruation (Albrecht et al., 1981).

How could oestradiol act within the corpus luteum? When oestradiol is infused locally through the ovary bearing the corpus luteum, progesterone secretion is suppressed and there is an associated rise in the concentration of $PGF_{2\alpha}$ in the ovarian vein (Auletta *et al.*, 1978). $PGF_{2\alpha}$ synergizes with oestrogen to induce luteal regression in the cynomolgus monkey suggesting that oestradiol may induce luteolysis by stimulating the production of $PGF_{2\alpha}$ (Shaikh and Klaiber, 1974). However, there are several other possibilities. Oestradiol *in vitro* inhibits 3β-ol dehydrogenase, the enzyme responsible for the conversion of pregnenolone to progesterone (Williams *et al.*, 1979). Moreover, in the cynomolgus monkey, bromocriptine markedly enhances the luteolytic effect of oestrogen (Castracane and Shaikh, 1980) (Figure 8). Although these authors concluded that this action of bromocryptine was due to suppression of prolactin secretion, it is likely that at the large dose used the secretion of LH was also suppressed. Thus the mechanism of action of oestrogen in the corpus luteum remains to be determined.

EFFECT OF BROMOCRIPTINE IN LUTEAL PHASE OF CYNOMOLGUS MONKEYS

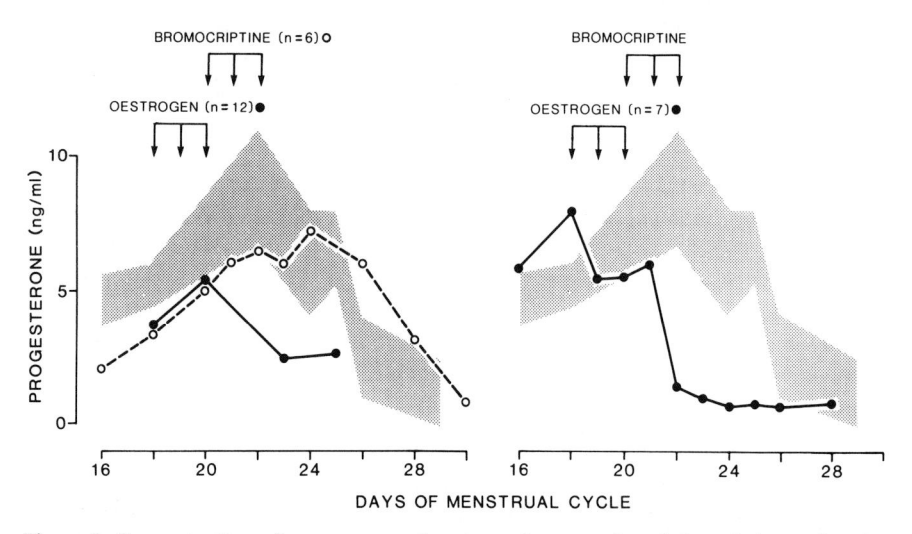

Figure 8. Concentration of progesterone in cynomolgus monkeys in luteal phase of cycle. The monkeys were injected with (*left panel*) either oestrogen (depoestradiol cyprionate, 49 μg on day 18 and 40 μg of oestradiol benzoate on days 19 and 20) or bromocriptine (1.0 mg/kg on days 20–22) separately, or (*right panel*) both oestrogen and bromocriptine. The combined treatment shortened the cycle to 23.5 ± 0.7 days. The shaded area is the mean progesterone concentration (±SEM) of 37 control cycles. (Redrawn from Castracane and Shaikh, 1980)

3.3. LHRH-like Peptide

When synthetic long-acting analogues of LHRH (luteinizing hormone releasing hormone) were tested, it was found that prolonged administration over several days resulted in suppression of gonadal activity in experimental animals (Corbin and Bex, 1980). This paradoxical effect of LHRH analogues has been confirmed in women, and its ability to suppress ovulation in women has been developed as a contraceptive (Nillius et al., 1978). Moreover, injection of synthetic analogues in the mid- but not in the early luteal phase resulted in premature suppression of progesterone secretion (Caspar and Yen, 1979).

The suppression of ovarian activity by these analogues is not due solely to their effect on the pituitary because they are effective in hypophysectomized animals treated with exogenous gonadotrophins. Studies in rats have demonstrated the presence of high-affinity low-capacity receptors in granulosa-lutein cells (Clayton et al., 1979). LHRH-like peptides will initially stimulate and then inhibit steroid secretion by ovarian cells by interfering with the synthesis of pregnenolone and the conversion of pregnenolone to progesterone as well as increasing the metabolism of progesterone by stimulating 20β-steroid dehydrogenase activity (Hsueh et al., 1983).

We have recently reported the presence of specific receptors for LHRH-like peptides in human luteal cells (Popkin et al., 1983). Their concentration varies significantly throughout the cycle with maximum levels being found during the mid-luteal phase (Bramley, Popkin, Fraser, Swanston and Baird, unpublished observations). Failure to demonstrate their presence in a previous study was probably due to the very high peptidase activity present in human luteal tissue (Clayton and Huhtaniemi, 1982). Progesterone production by human granulosa cells cultured in vitro was inhibited by a long-acting LHRH analogue (Tureck et al., 1982). These findings suggest that an LHRH-like peptide may play some physiological role in regulating the function of the corpus luteum although no such peptide has yet been identified in ovarian tissue.

3.4. Oxytocin and Relaxin

It has been known for some time that the corpus luteum can synthesize the polypeptide relaxin and more recently that its synthesis is stimulated by hCG (Quagliarello et al., 1980). However, among the more unexpected findings in recent years has been the discovery that the corpus luteum synthesizes oxytocin, a peptide which had hitherto been thought to be confined to the hypothalamus and posterior pituitary. During the process of extracting relaxin from the sheep corpus luteum it was noticed that there was a factor which consistently induced contraction rather than relaxation of smooth muscle in vitro (Wathes and Swann, 1982; Flint and Sheldrick, 1982). This factor has subsequently been identified as oxytocin and is present in human corpus luteum and is secreted into the ovarian vein (Baird and Flint, unpublished observations).

The factors which regulate the production of these peptides and their physiological role within the corpus luteum are still unknown. In sheep, immunization against oxytocin prolongs the luteal phase and release of oxytocin from the ovary is stimulated by injection of $PGF_{2\alpha}$ (Sheldrick et al., 1980). Moreover oxytocin will stimulate the release of $PGF_{2\alpha}$ from the uterus (Baird,

1978). However, in the sheep the luteolytic action of PGF$_{2\alpha}$ is not dependent on oxytocin release since luteolysis can be induced in a corpus luteum prolonged by hysterectomy in which situation the luteal content of oxytocin is very low (Sheldrick and Flint, 1983).

Do these observations have relevance to the regulation of ovarian activity in women? In low concentrations (4 mU/ml) oxytocin stimulates progesterone production from dispersed human luteal cells incubated *in vitro* (Tan et al., 1982). Higher concentrations inhibit both the basal and hCG-stimulated production of progesterone, suggesting that oxytocin, like PGF$_{2\alpha}$, may play a role in the intra-ovarian regression of luteal function.

4. Conclusions

Regression of the corpus luteum appears to depend on a balance between luteotrophic and luteolytic factors. It is virtually certain that LH is an important (perhaps the sole) part of the pituitary luteotrophic complex in most species including man. In pregnancy the luteotrophic effect of hCG secreted by the trophoblast can override any luteolytic influences. In the non-pregnant woman the corpus luteum is exposed to both luteotrophic and luteolytic factors and their relative effectiveness, which determines the function of the corpus luteum (Figure 1), changes as it ages (Table 2). Stimulatory factors include LH, PGE$_2$ and possibly prolactin, while PGF$_{2\alpha}$, oestradiol, LHRH-like peptides (and possibly oxytocin) are probably inhibitory. The production of these inhibitory factors may be influenced by the relative change in the concentration of steroids within the corpus luteum in favour of oestradiol. In species like the sheep, in which PGF$_{2\alpha}$ of uterine origin is the luteolytic factor, its synthesis and release are stimulated by oestrogen and falling levels of progesterone. Similar factors influence the synthesis of prostaglandins by the human uterus although in women they are involved in the mechanism of menstruation rather than luteal regression. Thus the life span of the corpus luteum in women is dependent on its

Table 2. Responsiveness of human corpus luteum to hormones at different stages of the luteal phase

Hormone	Phase: Day:	Early 14–19	Mid 19–24	Late 24–28
LH		—	SS	S
PGF$_{2\alpha}$		—	I	II
Oestrogen		—	I	II
LHRH		—	I	II
Oxytocin		?	I	?

S = stimulates progesterone secretion (SS = strongly).
I = inhibits progesterone secretion (II = strongly).

own secretory products — oestradiol and progesterone. By inhibiting the secretion of FSH and LH and reducing pituitary luteotrophic support the corpus luteum sows the seeds of its own destruction by rendering itself increasingly vulnerable to self-destruction by locally produced luteolytic factors.

References

Albrecht, E.D., Haskins, A.L., Hodgen, G.D., and Pepe, G.J. (1981). 'Luteal function in baboons with administration of the anti-estrogen ethomoxytriphetol (MER-25) throughout the luteal phase of the menstrual cycle.' *Biol. Reprod.*, **25**, 451–457.

Asch, R.H., Abou-Samra, M., Braunstein, G.D., and Pauerstein, C.J. (1982). 'Luteal function in hypophysectomized rhesus monkeys.' *J. Clin. Endocrinol. Metab.*, **55**, 153–161.

Auletta, F.J., Agins, H., and Scommegna, A. (1978). 'Prostaglandin F mediation of the inhibitory effect of estrogen on the corpus luteum of the rhesus monkey.' *Endocrinology*, **103**, 1183–1189.

Backstrom, C.T., McNeilly, A.S., Leask, R., and Baird, D.T. (1982). 'Pulsatile secretion of LH, FSH, prolactin, oestradiol and progesterone during the human menstrual cycle.' *Clin. Endocrinol.*, **17**, 29–42.

Baird, D.T. (1978). 'Local utero-ovarian relationships.' In *Control of Ovulation* (eds. D.B. Crighton, N.B. Hayes, G.R. Foxcroft and G.E. Lamming), pp. 217–233. Butterworth, London.

Baird, D.T., Baker, T.G., McNatty, K.P. and Neal, P. (1975). 'Relationship between the secretion of the corpus luteum and the length of the follicular phase of the human cycle.' *J. Reprod. Fertil.*, **45**, 611–619.

Balmeceda, J.P., Asch, R.H., Fernandez, E.O., Eddy, C.A., and Pauerstein, C.J. (1979). 'Prostaglandin production by rhesus monkey corpus luteum *in vitro*.' *Fertil. Steril.*, **31**, 214–216.

Beling, C.G., Marcus, S.L. and Markham, S.M. (1970). 'Functional activity of the corpus luteum following hysterectomy.' *J. Clin. Endocrinol. Metab.*, **30**, 30–39.

Carr, B.R., Sadler, R.K., Rochelle, D.B., Stalmach, M.A., MacDonald, P.C., and Simpson, E.R. (1981). 'Plasma lipoprotein regulation of progesterone biosynthesis by human corpus luteum tissue in organ culture.' *J. Clin. Endocrinol. Metab.*, **51**, 875–881.

Casper, R.F., and Yen, S.C.C. (1979). 'Induction of luteolysis in the human with a long-acting analogue of luteinizing hormone-releasing factor.' *Science*, **205**, 408–410.

Castracane, V.D., and Shaikh, A.A. (1980). 'Synergism of estrogen and bromocriptine in the induction of luteolysis in cynomolgus monkeys.' *J. Clin. Endocrinol. Metab.*, **51**, 1311–1315.

Challis, J.R.G., Calder, A.A., Dilley, A., Forster, S.S., Hillier, K., Hunter, D.J.S., McKenzie, I.Z., and Thorburn, G.D. (1976). 'Production of prostaglandin E and $F_{2\alpha}$ by corpus luteum, corpora albicantes and stroma from the human ovary.' *J. Endocrinol.*, **68**, 401–408.

Clayton, R.N., and Huhtaniemi, I.T. (1982). 'Absence of gonadotrophin-releasing hormone receptors in human gonadal tissue.' *Nature*, **297**, 56–59.

Clayton, R.N., Harwood, J.P. and Catt, K.T. (1979). 'Gonadotrophin-releasing hormone analogue binds to luteal cells and inhibits progesterone production.' *Nature*, **282**, 90–92.

Cooper, M.T., and Walpole, A.L. (1975). 'Practical application of prostaglandins in cervical husbandry.' In *Prostaglandins and Reproduction* (ed. S.M.M. Karim), pp. 309–328. MTP Press, Lancaster.

Corbin, A., and Bex, F.J. (1980). 'Luteinizing hormone releasing hormone and analogues. Conceptive and contraceptive potential.' In *Progress in Hormone Biochemistry and Pharmocology*, Vol. 1, (eds. M. Briggs and A. Corbin), pp. 227–297. MTP Press, Lancaster.

Corner, J.W., Jr. (1956). 'The histological dating of the corpus luteum of menstruation.' *Am. J. Anat.*, **98**, 377–401.

Dennefors, B.L., Sjogren, A., and Hamberger, L. (1982). 'Progesterone and adenosine 3′, 5′-monophosphate formation by isolated human corpus luteum at different ages: Influence of human chronic gonadotrophin and prostaglandins.' *J. Clin. Endocrinol. Metab.*, **55**, 102–107.

Flint, A.P.F., and Sheldrick, E.L. (1982). 'Ovarian secretion of oxytocin is stimulated by prostaglandin.' *Nature*, **297**, 587–588.

Gemmell, R.T., Stacy, B.D., and Thorburn, G.D. (1974). 'Ultrastructural study of secretory granulos in the corpus luteum of the sheep during the estrone cycle.' *Biol. Reprod.*, **11**, 447–462.

Gibori, G., and Miller, J. (1982). 'The ovary: Follicle development, ovulation and luteal function.' In *Biochemistry of Mammalian Reproduction* (eds. L.J.D. Zarrefeld and R.T. Chatterton), pp. 261–283. John Wiley, New York.

Gillim, S.W., Christensen, A.K., and McLennan, C.E. (1969). 'Fine structure of the human menstrual corpus luteum at a stage of maximum secretory activity.' *Am. J. Anat.*, **126**, 409–428.

Grinwich, D.L., Hichens, M. and Behrman, H.R. (1976). 'Control of the LH receptor by prolactin and prostaglandin $F_{2\alpha}$ in rat corpus luteum.' *Biol. Reprod.*, **14**, 212.

Hamberger, L., Nilsson, L., Dennefors, B., Kahn, I., and Sjogren, A. (1979). 'Cyclic AMP formation of human corpus luteum in response to LCF interference by $PGF_{2\alpha}$.' *Prostaglandins*, **17**, 615–621.

Hamberger, L., Kallfelt, B., Forshell, S., and Dukes, M. (1980). 'A luteolytic effect of a prostaglandin $F_{2\alpha}$ analogue in non-pregnant women?' *Prostaglandins*, **22**, 383–388.

Hanson, F.W., Powell, J.E., and Stevens, V.C. (1971). 'Effects of hCG and human pituitary LH on steroid secretion and functional life of the human corpus luteum.' *J. Clin. Endocrinol. Metab.*, **32**, 211–215.

Hearn, J.P. (1980). 'Primate models for early human pregnancy.' In *Animal Models in Human Reproduction* (eds. M. Serio and L. Martini), pp. 319–332. Raven Press, New York.

Henderson, K.M., and McNatty, K.P. (1975). 'A biochemical hypothesis to explain the mechanism of luteal regression.' *Prostaglandins*, **9**, 779–797.

Henderson, K.M., Scaramuzzi, R.J., and Baird, D.T. (1977). 'Simultaneous infusion of prostaglandin E_2 antagonizes the luteolytic action of prostaglandin $F_{2\alpha}$ *in vivo*.' *J. Endocrinol.*, **72**, 379–383.

Hoffmann, F. (1960). 'Untersuchunger uber die hormonale Beeinflussung der Levensdauer des Corpus Luteum in Zyklus der Frau.' *Geburtshilfe Frauenheilkd.*, **20**, 1153–1159.

Horton, E.W., and Poyser, N.L. (1976). 'Uterine luteolytic hormone: A physiological role for prostaglandin $F_{2\alpha}$.' *Physiol. Rev.*, **56**, 595–651.

Hsueh, A.J.W., Jones, P.B.C., Adashi, E.Y., Wang, C., Zhlang, L.Z., and Welsh, T.H.J. (1983). 'Intra ovarian mechanisms in the hormonal control of granulosa cell differentiation in rats.' *J. Reprod. Fertil.*, **69**, 325–342.

Hunter, M.G. (1980). 'Studies of the corpus luteum *in vitro*.' Ph.D. Thesis, University of Edinburgh.

Johansson, E.D.B., and Gemzell, C. (1971). 'Plasma levels of progesterone during the luteal phase in normal women treated with synthetic oestrogens (RS 2874, F 6103 and ethinyl oestradiol).' *Acta Endocrinol. (Copenhagen)*, **68**, 551–560.

Karsch, F.J., and Sutton, G.P. (1976). 'An intraovarian site for the luteolytic action of estrogen in the rhesus monkey.' *Endocrinology*, **98**, 553–561.

Keyes, P.L., Gadsby, J.E., Yuh, K.M., and Bill, C.H. (1983). 'The corpus luteum.' In *International Review of Physiology*, Vol. 27 (ed R.O. Greep), pp. 57–97. University Park Press, Baltimore.

Kirton, K.T. (1975). 'Prostaglandins and reproduction in sub-human primates.' In *Prostaglandins and Reproduction* (ed. S.M.M. Karim), pp. 229–240. MTP Press, Lancaster.

Knobil, E. (1973). 'On the regulation of the primate corpus luteum.' *Biol. Reprod.*, **8**, 246–258.

Korda, A.R., Shutt, D.A., Smith, I.D., Shearman, R.P. and Lyneham, R.C. (1975). 'Assessment of possible luteolytic effect of intra ovarian injection of prostaglandin $F_{2\alpha}$ in the human.' *Prostaglandins*, **9**, 443–449.

McCracken, J.A., Baird, D.T., and Godine, J.R. (1971). 'Factors affecting the secretion of steroids from the transplanted ovary in the sheep.' *Recent Prog. Horm. Res.*, **27**, 537–582.

McNatty, K.P., Sawers, R.S., and McNeilly, A.S. (1974). 'A possible role for prolactin in control of steroid secretion by the human Graafian follicle.' *Nature*, **250**, 653–655.

McNatty, K.P., Henderson, K.M., and Sawers, R.S. (1975). 'Effects of prostaglandins $F_{2\alpha}$ and E_2 on the production of progesterone by human granulosa cells in tissue culture.' *J. Endocrinol.*, **67**, 231–240.

McNeilly, A.S., Kerin, J., Swanston, I.A., Bramley, T.A., and Baird, D.T. (1980). 'Changes in the binding of human chorionic gonadotrophin/luteinizing hormone, follicle stimulating hormone and prolactin to human corpus luteum during the menstrual cycle and pregnancy.' *J. Endocrinol.*, **87**, 315–325.

Manaugh, L.C., and Novy, M.J. (1976). 'Effect of indomethacin on corpus luteum function and pregnancy in rhesus monkeys.' *Fertil. Steril.*, **27**, 588.

Marsh, J.M. (1976). 'The role of cyclic AMP in gonadol steroidogenesis.' *Biol. Reprod.*, **14**, 30–53.

Midgley, A.R., Jr., and Jaffe, R. (1968). 'Regulation of human gonadotrophins: IV. Correlation of serum concentrations of follicle stimulating and luteinizing hormones during the menstrual cycle.' *J. Clin. Endocrinol. Metab.*, **28**, 1699–1703.

Mougdal, N.R., MacDonald, G.J., and Greep, R.O. (1971). 'The effect of h.C.G. antiserum on ovulation and corpus luteum formation in the monkey.' *J. Clin. Endocrinol. Metab.*, **32**, 579–581.

Mougdal, N.R., MacDonald, G.J., and Greep, R.O. (1972). 'Role of endogenous primate LH in maintaining corpus luteum function of the monkey.' *J. Clin. Endocrinol. Metab.*, **35**, 113–116.

Nillius, S.J., Bergquist, C., and Wide, L. (1978). 'Inhibition of ovulation in women by chronic treatment with a stimulatory LRH analogue – a new approach to birth control?' *Contraception*, **17**, 537–545.

Patwardhan, V.V., and Lanthier, A. (1980). 'Concentration of prostaglandins PGE and PGF, estrone, estradiol and progesterone in human corpus luteum.' *Prostaglandins*, **20**, 963–969.

Poindexter, A.N., Buttram, V.C., Besch, P.K., and Smith, R.G. (1979). 'Prolactin receptors in the ovary.' *Fertil. Steril.*, **31**, 273–277.

Popkin, R., Bramley, T.A., Currie, A., Shaw, R.W., Baird, D.T., and Fraser, H.M. (1983). 'Specific binding of luteinizing hormone releasing hormone to human luteal tissue.' *Biochem. Biophys. Res. Commun.*, **114**, 750–756.

Powell, W.S., Hammarström, S., Samuelsson, B. and Sjöberg, B. (1974). 'Prostaglandin $F_{2\alpha}$ receptor in human corpus luteum.' *Lancet*, **i**, 1120.

Quagliarello, J., Smith, L., Steinetz, B., and Weiss, G. (1980). 'Induction of relaxin secretion in non-pregnant women by human chorionic gonadotrophin.' *J. Clin. Endocrinol. Metab.*, **51**, 74.

Rajaniemi, H.J., Ronnberg, L., Kauppila, A., Ylostalo, P., Jalkanen, M., Saastamoinen, J., Selander, K., Pystynen, P., and Vihko, R. (1981). 'Luteinizing hormone receptors in human ovarian follicles and corpus luteum during the menstrual cycle and pregnancy.' *J. Clin. Endocrinol. Metab.*, **108**, 307–313.

Rothchild, I. (1981). 'The regulation of the mammalian corpus luteum.' *Recent Prog. in Horm. Res.*, **37**, 183.

Santen, R.J., and Bardin, C.W. (1973). 'Episodic LH secretion in man: Pulse analysis, clinical interpretation and physiologic mechanisms.' *J. Clin. Invest.*, **52**, 2617–2682.

Schoonmaker, J.N., Victery, W., and Karsch, F.J. (1981). 'A receptive period for estradiol induced luteolysis in the rhesus monkey.' *Endocrinology*, **108**, 1874–1877.

Schoonmaker, J.N., Bergman, K.S., Steiner, R.A., and Karsch, F.J. (1982). 'Estradiol induced luteal regression in the rhesus monkey: Evidence for an extra-ovarian site of action.' *Endocrinology*, **110**, 1708–1715.

Shaikh, A.A., and Klaiber, E.L. (1974). 'Effects of sequential treatment with estradiol and $PGF_{2\alpha}$ in the length of the primate menstrual cycle.' *Prostaglandins*, **6**, 253–262.

Sheldrick, E.L., and Flint, A.P.F. (1983). 'Regression of the corpus luteum in sheep in response to cloprostenol is not affected by loss of luteal oxytocin after hysterectomy.' *J. Reprod. Fertil.*, **68**, 155–160.

Sheldrick, E.L., Mitchell, M.D., and Flint, A.P.F. (1980). 'Delayed luteal regression in ewes immunized against oxytocin.' *J. Reprod. Fertil.*, **59**, 37–42.

Shutt, D.A., Clarke, A.H., Fraser, I.S., Goh, P., McMahon, G.R., Saunders, D.M., and Shearman, R.O. (1976). 'Changes in concentration of prostaglandin $F_{2\alpha}$ and steroids in human corpus luteum in relation to growth of the corpus luteum and luteolysis.' *J. Endocrinol.*, **71**, 453–454.

Swanston, I.A., McNatty, K.P., and Baird, D.T. (1977). 'Concentration of prostaglandin $F_{2\alpha}$ and steroids in the human corpus luteum.' *J. Endocrinol.*, **73**, 115–122.

Tan, G.J.S., Tweedale, R., and Biggs, J.S.G. (1982). 'Oxytocin may play a role in the control of the human corpus luteum.' *J. Endocrinol.*, **95**, 65–70.

Tureck, R.W., Mastrionni, L., Blasco, L., and Strauss, J.F. (1982). 'Inhibition of human granulosa cell progesterone secretion by a gonadotrophin-releasing hormone agonist.' *J. Clin. Endocrinol. Metab.*, **54**, 1078–1080.

Van de Wiele, R.L., Bogumil, J., Dyrenfurth, I., Ferin, M., Jewelewicz, R., Warren, M., Rizkallah, T., and Mikhail, G. (1970). 'Mechanisms regulating the menstrual cycle in women.' *Recent Prog. Horm. Res.*, **26**, 63–103.

Wathes, D.C., and Swann, R. (1982). 'Is oxytocin an ovarian hormone?' *Nature*, **297**, 225–227.

Wentz, A.C., and Jones, G.S. (1973). 'Transient luteolytic effect of prostaglandin $F_{2\alpha}$ in the human.' *Obstet. Gynecol.*, **42**, 172–181.

Williams, M.T., Roth, M.S., Marsh, J.M., and Lemaire, W.J. (1979). 'Inhibition of human chorionic gonadotrophin induced progesterone synthesis by estradiol in isolated human luteal cell.' *J. Clin. Endocrinol. Metab.*, **48**, 437–440.

The Luteal Phase
Edited by S.L. Jeffcoate
© 1985 John Wiley & Sons Ltd.

CHAPTER 3

Uterine responses to the corpus luteum

C.A. FINN
*Department of Veterinary Physiology,
Leahurst,
Neston,
Wirral, UK*

Introduction

The corpus luteum influences the uterus primarily through its secretion of progesterone together with low levels of oestrogen. However, it clearly cannot develop without there first having been a follicle which also secretes high levels of oestrogen; so in any physiological sense it is pointless to discuss the responses of the uterus to progesterone without considering also the action of follicular oestrogen. As far as the uterus is concerned the two hormones act together and their action is closely interrelated. Follicular oestrogen is also important because it is the signal that sets off the train of events which not only results in the formation of the corpus luteum but also sets the gamete on its way, and at the same time starts acting on the uterus and uterine tubes ensuring that uterine changes are co-ordinated with insemination, fertilization and transport of the ovum. This means that during the luteal phase the uterus is prepared for implantation at the same time as the blastocyst is mature and present in the uterus and ready to attach. Oestrogen secreted during the follicular phase thus plays a pivotal role in controlling implantation into the uterus.

The ovarian hormones influence both parts of the uterus. During the luteal phase the main influence is probably on the endometrium with the myometrium becoming more significant during pregnancy and parturition. Nevertheless the hormones may have some effect on the myometrium during menstruation, which although, strictly speaking, is not part of the luteal phase, is a direct result of it. The menstrual products come from the endometrium but their removal from the uterus involves myometrial activity. This, however, occurs after the breakdown of the corpus luteum so it is unlikely that secretions of the corpus luteum play much part in its control except maybe by inhibiting muscular activity during the luteal phase. The main emphasis in this chapter, therefore, will be on the influence of oestrogen and progesterone on the endometrium.

I shall concentrate mainly on the cellular aspects because it is important that these are clearly defined before biochemical changes are analysed. Gross biochemical estimations of uterine extracts that do not take account of changing cell populations can be very misleading.

There is also the problem of species variation. The method of implantation and the cellular changes in the endometrium in preparation for implantation vary considerably from species to species and it is therefore not meaningful to extrapolate directly from one species to another. This is particularly relevant in this book, where the aim is to provide information of interest to human clinical medicine. There has been much exerimental work on the uterus carried out on laboratory rodents whereas experimental work on the human uterus is very limited. On the other hand, there is quite a lot of observational data on the human endometrium. Whilst it would be wrong to try to apply experimental findings on the mouse directly to the human it would equally be silly to discard all the rodent data. In this paper an attempt is made to show how far the cellular changes that have been worked out in rodents may be relevant to the human,

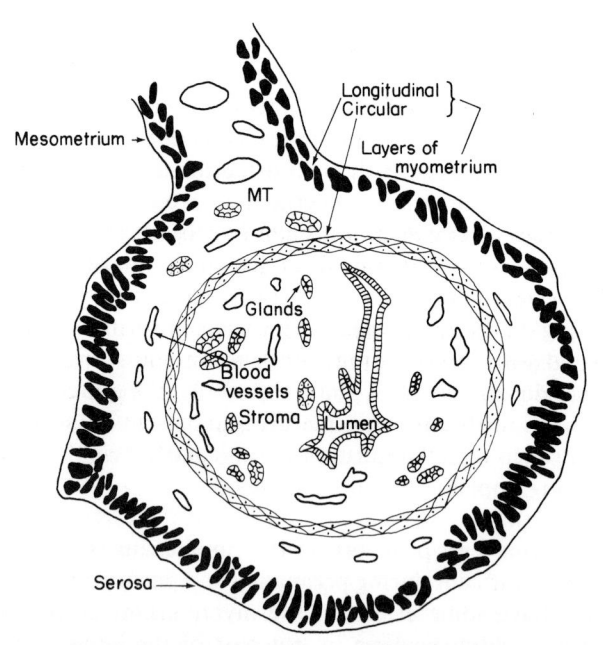

Figure 1. Schematized cross-section of a uterus showing the main tissues (based on the mouse uterus). In the human uterus there are three muscle layers. MT = area of mesometrial triangle in which metrial gland forms

and to show how it may be possible to adapt the changes in mice to give a better model for the human.

As discussed earlier, the uterine changes in the luteal phase which culminate in implantation or in menstruation, start during the follicular phase under the influence of oestradiol. The endometrium is made up of three main tissues — luminal epithelium, glandular epithelium and stroma (Figure 1). The latter contains the blood and lymph vessels and several different cell types. Most important among these is the undifferentiated fibroblast which has the potential to differentiate into the decidual cell. There are also varying numbers of white blood cells and mast cells.

From work on the mouse it is apparent that under the influence of the ovarian hormones each of the tissues in the days leading up to implantation first undergoes a burst of cell proliferation, then a period of differentiation, followed by cell death. In order to analyse the hormonal control of these processes ovariectomized mice can be injected with exogenous hormones and then histological or electron microscope sections of the uteri examined at various times after the hormone injections (Martin and Finn, 1971). Such experiments, however, will only be meaningful if the dosage and timing of hormone injections mimic as far as possible the pattern of hormone secretion during the cycle or early pregnancy. It is particularly important that oestrogens are not given in large doses.

1. Cell Division

If ovariectomized mice are given daily injections of 100 ng of oestradiol and killed on successive days, it is found that the first response is for the luminal epithelial cells to undergo mitosis. This is maximal on the first two days of the oestrogen injections and is best demonstrated if colchicine is given 2 h before autopsy (as shown in Table 1). Surprisingly there is little glandular mitosis at this time, but if one waits for a further 2 days then there is a wave of division in the glands. This second response will also occur if the injections of oestradiol are stopped (Table 2); for instance, if oestrogen is given for 2 days and the animals are killed on subsequent days then glandular mitosis is maximal after a gap of 2 days (Finn and Martin, 1973). Thus for the gland cells to divide it is necessary for a more protracted period of oestrogen stimulation or for oestrogen treatment to be stopped. The reason for this is not known. The glands originate as downgrowths of the luminal epithelium so it is rather surprising that they respond differently.

In women it is apparent that oestrogens cause division of the epithelial cells although it is not known whether the responses in the lumen and glands occur at different times as in the mouse. The human endometrium contains a much higher proportion of glandular to luminal tissue than the mouse and in biopsies it is not always easy to differentiate the two, and of course it is not possible to administer colchicine before taking the biopsies. It is also found in mice that

Table 1. Luminal and glandular mitosis after continuous daily injections of 100 ng of oestradiol-17β

		No. of cells in mitosis † (mean + SEM)	
Treatment*	n	Lumen	Glands
1	5	56.2 ± 5.7	5.8 ± 1.7
2	10	50.7 ± 5.7	10.1 ± 1.5
3	10	36.6 ± 8.3	38.7 ± 5.6
4	9	29.8 ± 8.4	23.0 ± 3.2
5	9	12.4 ± 2.9	16.8 ± 2.7

* Number of days of oestradiol treatment preceding autopsy.
† Counts represent the total number of cells undergoing mitosis in a cross-section of the uterus taken at random.

Table 2. Mitosis after two daily injections of oestradiol (five mice per group)

	Number of cells in mitosis (mean ± SEM)	
Treatment*	Lumen	Glands
EK	69.6 ± 9.8	5.6 ± 1.4
EEK	85.6 ± 16.7	12.0 ± 2.7
EE – K	0	9.2 ± 3.5
EE – – K	4.0 ± 0.9	32.6 ± 5.7
EE – – – K	10.6 ± 1.8	20.8 ± 3.4

* E = 100 ng of oestradiol; – = no injection; K = day of autopsy.

Table 3. Uterine mitosis 24 h after a single injection of hormone(s) on the days after oestrogen priming and with no priming (day of ovariectomy = day 1)

Treatment on day								No. of mice	No. of mitoses (mean ± SE)		
3	4	5	6	7	8	9	10	11		Luminal	Glandular	Stromal
E	E	–	–	P	K	–	–	–	15	1.9 ± 0.5	3.3 ± 0.7	1.8 ± 0.5
E	E	E	–	–	P	K	–	–	15	0.7 ± 0.2	1.9 ± 0.7	20.1 ± 5.4
E	E	E	–	–	–	P	K	–	15	2.3 ± 0.8	5.9 ± 1.7	16.8 ± 5.3
E	E	E	–	–	–	–	P	K	10	4.4 ± 1.2	13.7 ± 3.3	2.5 ± 0.8
–	–	–	–	–	–	P	K	–	14	6.4 ± 1.5	7.3 ± 1.2	0.9 ± 0.3

E = 0.1 μg of oestradiol; P = 1 mg of progesterone; K = killing.

proliferation of the epithelial cells does not only occur in response to hormones; it also follows damage to the endometrium. Thus in mice if the luminal epithelium is squeezed out of the uterus, it very rapidly reforms from, presumably, remaining glandular cells (Lejeune et al., 1981). Probably a similar process occurs after menstruation in women.

If progesterone is given to mice following oestrogen treatment further changes occur. The stromal cells which did not respond to oestrogen now undergo mitosis. However, this only occurs if a gap of 2 or 3 days is left between the oestrogen and progesterone injections. The results of the experiment shown in Table 3 show how the mitosis in the stroma varies with variation in the gap (Finn and Martin, 1970). Presumably during this interval the follicular oestrogens are bringing about changes in the stromal cells preparing them to respond to the progesterone or progesterone plus luteal oestrogen. Progesterone given without oestrogen priming will not cause stromal mitosis although if it is given alone for at least 4 days mitosis can be induced in the stroma by giving a small dose of oestrogen in addition to the progesterone. This schedule of hormone injections mimics the pattern of hormone secretion in early pregnancy in the mouse and women where there is first secretion of high levels of oestrogen during pro-oestrus, then a period when the corpus luteum is developing when no hormone is present before progesterone secretion starts. It confirms what was said earlier about the interrelationship between the two hormones. Whether stromal mitosis in women is dependent on progesterone preceded by oestrogen is not known for certain although it certainly occurs during the luteal phase (Noyes et al., 1950) when such conditions exist. Thus we can see that the pattern of hormone changes brought about by the formation of the follicle, ovulation and the formation of the corpus luteum bring about a well-defined pattern of cell division in the endometrium.

2. Cell Differentiation

2.1. Epithelial Cells

Progesterone, together with the small quantity of oestradiol secreted during the luteal phase, is largely responsible for the differentiation of the endometrial cells which follows their proliferation. Each of the tissues has a characteristic pattern of differentiation in preparation for the role each has to play in implantation. In mice the luminal epithelial cells are the first to differentiate. This takes place in two stages. In the first, which we call the first stage of closure, the opposing surfaces of the lumen come together so that the microvilli on opposing surfaces interdigitate (Figure 2). This is followed by the second stage of closure when the microvilli disappear and opposing surfaces are very closely applied so that the lumen is almost obliterated (Figure 3) (Pollard and Finn, 1972). This differentiation is associated with the attachment of the blastocyst, which in

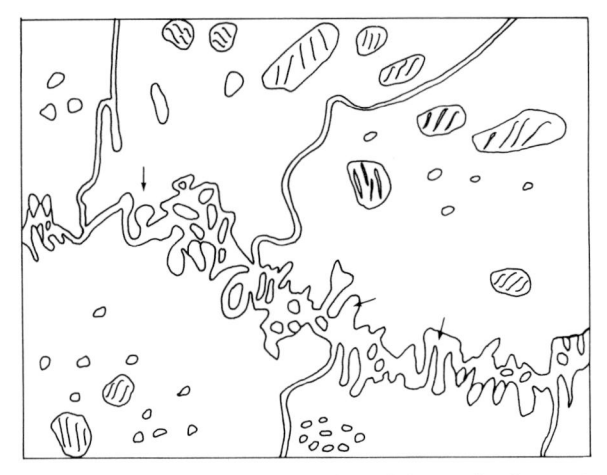

Figure 2. Line drawing showing the configuration of the uterine lumen (arrowed) in the first stage of closure. Microvilli from opposing surfaces are interlocked

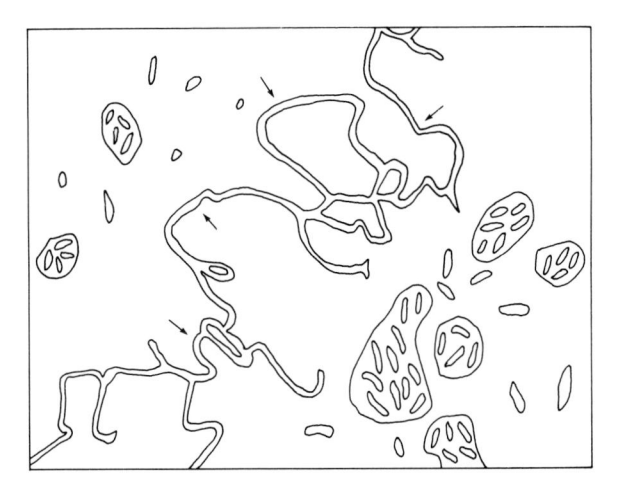

Figure 3. Line drawing showing the configuration of the uterine lumen (arrowed) in the second stage of closure. Microvilli are no longer discernible and membranes of opposing epithelial cells are in very close contact

mice is the first stage of implantation (Figure 4). The first stage can be produced in ovariectomized mice by giving progesterone for several days [less if priming oestrogen is given (Pollard and Finn, 1972)], whilst for the second stage a small quantity of oestrogen on top of the progesterone is necessary. Once induced, the second stage can only be held for a short period, after which a third stage follows in which attachment cannot occur. This gives a very precise timing to implantation in mice with the uterus only sensitive to implantation for a very

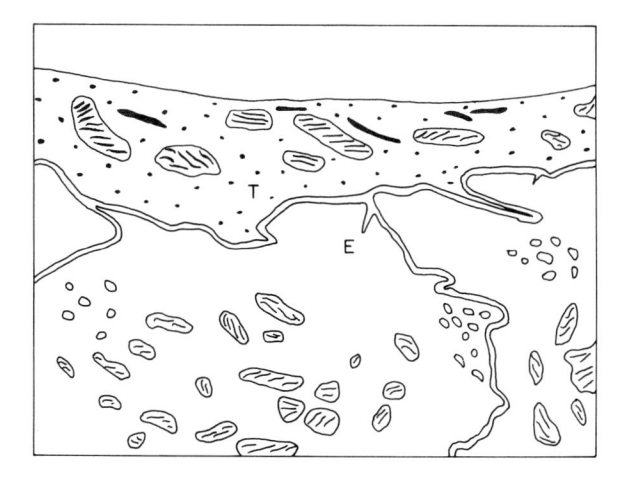

Figure 4. Line drawing showing the relationship of the trophoblast (T) to the surface of the uterine epithelial cells (E) after attachment of the blastocyst

short period associated with differentiation of the endometrium (Finn and Pollard, 1973).

To demonstrate the closure reaction it is necessary for the uterus to be very carefully fixed, ideally using the fixatives used in electron microscopy (glutaraldehyde by vascular perfusion and osmic acid). Furthermore differentiation between the first and second stages is only possible with magnifications obtained with the electron microscope. Biopsy sections taken from the inside of the uterus would obviously not demonstrate closure and sections through whole human uteri under the correct hormone conditions and adequately fixed are difficult to come by. Thus we do not know to what extent the human uterus closes under the influence of progesterone. Circumstantial evidence suggests that differentiation of the luminal surface of the endometrium may not be so important in women as in rodents. The mode of implantation in humans appears to be somewhat different from in the mouse (although it must be admitted that nobody has actually seen the earliest stages in women). In the mouse the first stage of implantation is for the trophoblast to become attached to the surface of the uterine epithelium; the latter then breaks down by a process of programmed cell death (Finn and Bredl, 1973) so that the blastocyst passively finds itself inside the stroma. In the human, the trophoblast appears to burrow actively between the luminal epithelial cells. The only common non-primate showing a similar type of implantation is the guinea pig. The uterus of the guinea pig has been studied with the electron microscope (Green, 1979) and, although the first stage of closure was demonstrated, the second stage, which represents the final differentiation of the cell surface, was not observed. Indeed as implantation does not appear to need close attachment between the trophoblast and the luminal surface of the endometrium there may not be a necessity for such differentiation of the surface.

It is probable that the luminal epithelium of the human uterus plays a more passive role in implantation with the blastocyst being more active. If this were so it would probably mean that the timing of implantation might not be so critical in women. This is obviously very important from the point of view of embryo transfer and more information is needed.

2.2. Uterine Glands

Differentiation of the uterine glands involves preparing them for the synthesis of a periodic acid–Schiff-positive substance which is then secreted into the gland lumen. This can be clearly seen in stained sections. In the mouse, gland secretion occurs in response to both progesterone and oestrogen (Finn and Martin, 1976). Very little gland secretion is observed when progesterone alone is given to ovariectomized mice, but the addition of a small quantity of oestrogen induces copious secretion. Gland secretion is very prominent during the luteal phase in women and is dependent on progesterone; whether it is also affected by the small levels of oestradiol during the luteal phase is not known. Human gland cells differ from the mouse in containing masses of glycogen in their cytoplasm during the late luteal phase as well as a characteristic nucleolar system (Dubransky and Pohlman, 1961) and giant mitrochondria (Gompel, 1964). The significance of these is unknown but they appear to be unique to the human.

2.3. Stroma

Differentiation of the stroma is a complex and interesting process. As already mentioned, in both rodents and women the blastocyst during implantation comes to life within the stroma. The trophoblast is a very invasive tissue (Kirby, 1960) so this presents the uterus with a problem – ordinary connective tissue would be rapidly destroyed. Furthermore, in these species the blastocyst remains small and a very efficient vascular supply must therefore be prepared. Differentiation consists of growth of the stromal blood vessels (angiogenesis) with increase in their permeability and the transformation of the stromal fibroblasts into large decidual cells. The presence of the decidua was first shown in women more than 200 years ago by William Hunter (1794). What Hunter described was a macroscopically visible membrane or tissue including blood vessels and various cells. It is important to remember that changes in the endometrial blood vessels are a very important part of decidualization. Since Hunter's time the decidua has been most widely studied in rodents. In fact apart from women only a few species actually show decidual transformation of the stroma. As would be expected the decidua appears to develop only in those species in which during implantation the blastocyst migrates inside the stroma, for example some primates and most rodents.

Decidualization is dependent on progesterone (Nelson and Pfiffner, 1930). There is, however, a very important difference between decidualization in the rodent and human. In the mouse, although decidual transformation is dependent on progesterone it will not occur even when the hormone conditions are correct unless a stimulus is given to the surface of the uterine epithelium by the trophoblast, although this can under experimental conditions be replaced by an artificial stimulus such as a drop of oil (Finn and Keen, 1962). Thus if ovariectomized mice are treated first with high doses of oestrogen, then after an interval of 3 days they are given progesterone plus a very small dose of oestradiol for 3 days, the uterus becomes sensitive and will respond to an intraluminal injection of oil by undergoing the decidual reaction (Finn and Martin, 1972). This mimics the conditions of early pregnancy in the mouse, with the oil replacing the blastocyst. The first sign of transformation is that the stromal blood vessels become more permeable to large molecules. This can easily be demonstrated by injecting a dye such as pontamine sky blue intravenously, and then killing the animal after about 15 min. Decidualizing areas in the uterus stain blue (Psychoyos, 1960). When sections are cut through these areas the first change seen is oedema of the stroma presumably due to the increased passage of fluid from the permeable blood vessels (Krehbiel, 1937). A little later the stromal fibroblasts nearest the lumen are seen to be enlarged, often with two or more nuclei. The nuclei are very large due to polyploidy (Sachs and Shelesnyak, 1955). Another important characteristic of these decidual cells is that they are joined together by specialized gap junctions thus forming a continuous mass of tissue (Finn and Lawn, 1967). They also synthesize specific enzymes such as alkaline phosphatase (Finn and Hinchliffe, 1964) and ornithine decarboxylase (Collawn et al., 1981) and contain glycogen granules (Krehbiel, 1937). A little later another cell type also appears in the stroma. This is a granulated cell which is thought to originate from the blood monocyte (Selye and McKeown, 1935). These eventually settle in the metrial gland but early on they are found scattered throughout the stroma. Normally decidualization in the mouse and other rodents only occurs during pregnancy; it does not occur during the pseudopregnant cycle which is equivalent to the normal ovarian cycle in other animals.

In women, on the other hand, differentiation of the stroma starts during the luteal phase of the normal mentrual cycle. As with the mouse the first change is in the stromal blood vessels which grow considerably during the luteal phase. It has not been possible to demonstrate definitely a change in permeability using a dye, as in the mouse, but in sections the stroma can be seen to become oedematous indicating probably that the capillaries have become more permeable. Following this some of the stromal fibroblasts, especially those around the blood vessels, transform into decidual cells. These are large cells which contain alkaline phosphatase (Wilson, 1969) and are joined by gap junctions (Lawn et al., 1971). Granulated cells (K cells) also appear in the stroma during the luteal phase (Hamperl, 1955). These are probably analogous to the metrial gland cells in the rodent. If implantation occurs and the corpus luteum is maintained or if exogenous progestin is given (Eichner *et al.*, 1951), further development of the decidual cells and blood vessels takes place to form the decidual membrane which is shed with the placenta.

As in the rodent, the continued development and survival of the decidua is completely dependent on progesterone, so that if a blastocyst is not present and the corpus luteum degenerates then the decidual changes in the stroma are not maintained. The consequent breakdown of the decidua results in bleeding from exposed blood vessels. Thus it appears that menstruation has come about because of the evolution towards greatly increased vascular and cellular differentiation of the stroma during the cycle in women and some primates. This idea was originally put forward by Grosser in 1910. He stated that the predecidual changes in the endometrium represent 'a preparation for the reception of a fertilized ovum' and that 'they are physiologically the most important part of the entire cycle, while menstruation itself is only a secondary process, a degeneration of the mucous membrane which from a failure of pregnancy has not been able to fulfil its purpose.' This may sound obvious to us now but at that time such eminent scientists as Marshall (1910) and Bryce and Teacher (1908) thought that menstruation was equivalent to the bleeding from the uterus seen in some animals during heat. It seems that other animals do not menstruate because they do not decidualize during the cycle, although some do show bleeding during times when there are high levels of oestrogen acting on the blood vessels as during pro-oestrus.

If it is correct that menstruation follows because of the lack of progestational support for the decidualized stroma, then it should be possible to induce a similar condition to menstruation in ovariectomized mice by using the appropriate hormones to prepare the endometrium, inducing decidualization with an artificial stimulus and then stopping the hormone injections.

To test this hypothesis ovariectomized mice were given hormones on the schedule shown in Table 4 (Finn and Pope, 1984). These hormones bring about the cellular changes already described and make the uterus sensitive to a

Table 4. Schedule of hormone injections administered to ovariectomized mice to prepare the uterus to respond to a decidual stimulus

Day 1	100 ng of oestradiol
Day 2	100 ng of oestradiol
Day 3	
Day 4	No treatment
Day 5	
Day 6	10 ng of oestradiol plus 500 μg of progesterone
Day 7	10 ng of oestradiol plus 500 μg of progesterone
Day 8	10 ng of oestradiol plus 500 μg of progesterone; 0.02 ml of arachis oil injected into uterine lumen 4–6 h after hormone injections

decidual stimulus. To induce decidualization 0.02 ml of arachis oil was injected into the uterine lumen on day 8, 4 h after the injections of progesterone and oestrogen. Control animals were injected with a long-acting progestin (medroxyprogesterone acetate) after the oil injections and were killed at 31, 45 and 60 h after the oil injection, and the uteri sectioned. Normal decidual reactions were obtained in these animals − normal decidual cells and blood vessels (Figure 5). Experimental animals were given no further progestin after decidual induction and were killed at various times over the following 84 h. A definite pattern of changes took place in the uteri which can be summarized as follows.

At 31 h a normal decidual reaction could be seen, indistinguishable from the controls. By 45 h clear differences appeared in about half of the treated animals (4/8) and by 48 h all the animals showed changes. The first change appeared in the blood vessels, especially those around the lumen. They were dilated and filled with swollen red blood cells (Figure 6) and sometimes platelets. Very soon

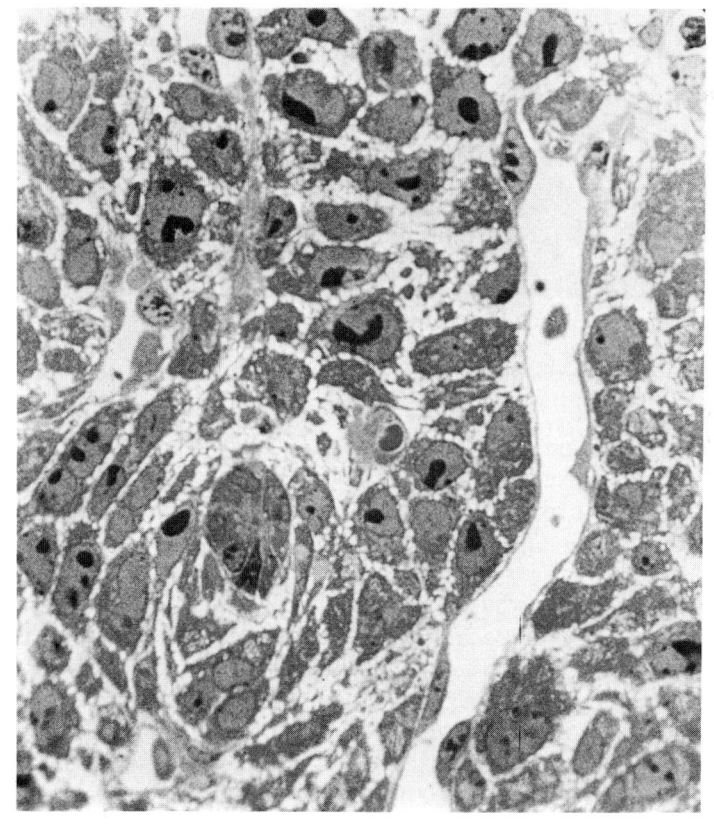

Figure 5. Cross-section of uterus from control animal which had been injected with oil 57 h before autopsy, showing normal decidual cells and muscle. × 700. (From Finn and Pope, 1984)

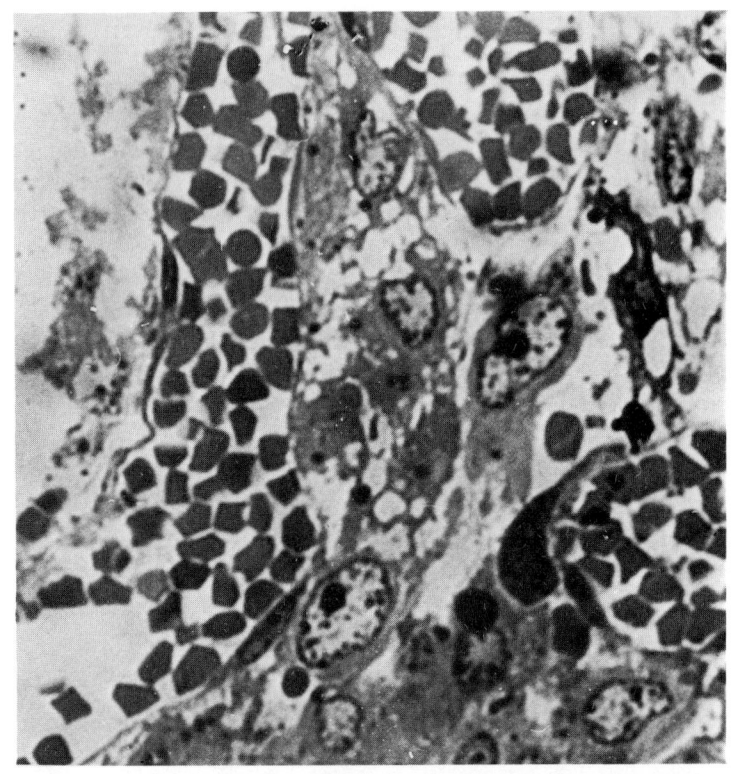

Figure 6. Cross-section of uterus from experimental animal which had been injected with oil but had received no progesterone for 48 h before autopsy, showing congested blood vessels and decidual cells with granulated nuclei. × 200. (From Finn and Pope, 1984)

after this the decidual cells showed signs of cell death. The nuclei took on a granulated appearance, the first sign of apoptosis.

At a later stage, from 55 h, discontinuities appear in the walls of the congested blood vessels so that blood cells are found extravascularly (Figure 7). These changes occur in the area of stoma around the lumen so that two zones become clearly differentiated (Figure 8) – an outer rim of normal looking stromal fibroblasts, glands and blood vessels and an inner degenerating zone containing masses of blood cells and degenerating decidual cells. Amongst the cells of the outer rim are large numbers of leucocytes presumably migrating to the inner zone (Figure 9). The luminal epithelium is incomplete and blood cells are present in the lumen; in about 50% of cases blood is seen macroscopically in the vagina. Later, from about 70 h, the inner core is seen to be detaching from the outer rim, and by 85 h it had shelled off in all animals. The differentiation of the stroma into an inner zone which breaks down and an outer zone which remains intact is very similar to the condition in the human uterus, except that in

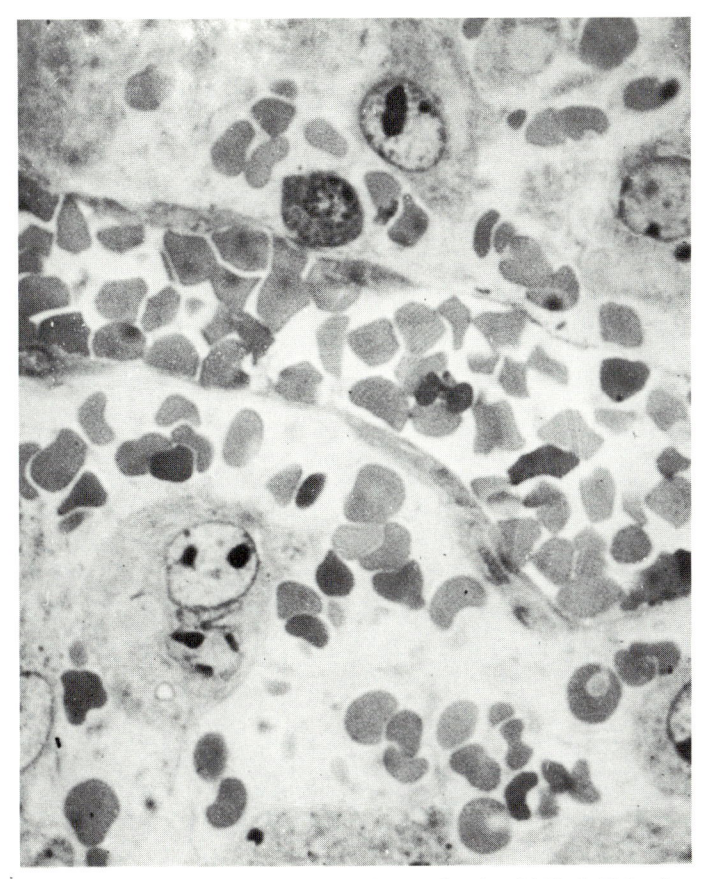

Figure 7. Cross-section of uterus from experimental animal killed 48 h after the progesterone injection, showing extravasation of blood cells into stroma. × 1000

the mouse under normal physiological conditions it is not possible to differentiate the stroma into distinct areas.

From these studies it is apparent that following the removal of progesterone the first and most prominent endometrial change occurs in the blood vessels of the stroma. This is similar to menstruation in women and raises some interesting questions. Why should the blood vessels become congested and break down? Are newly formed blood vessels particularly sensitive to changes in hormone environment or is the breakdown secondary to the congestion and if so what causes this? Again one might ask why the decidual cells die. Is it secondary to the vascular effect or is it programmed cell death? In mice that have been treated on exactly the same hormone schedule but not had oil injected into their uteri there is no sign of any such changes upon withdrawal of the progesterone, so the vascular effects must be a consequence of the initial changes induced by

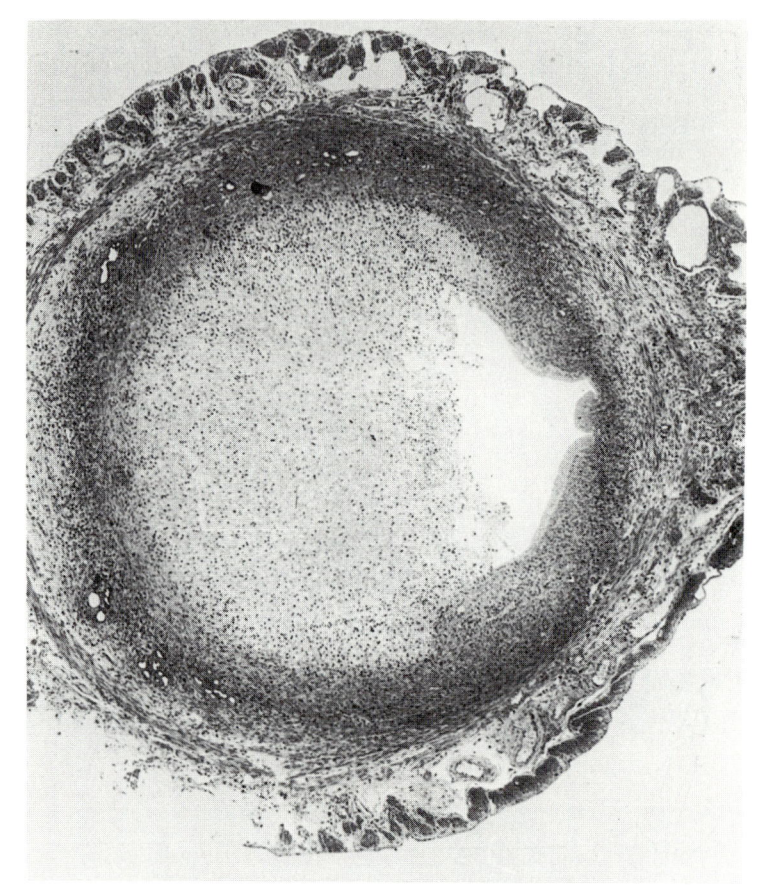

Figure 8. Cross-section of uterus from an animal killed at 56 h after last progesterone injection, showing central degenerating area of stroma. × 40 (From Finn and Pope, 1984)

the oil acting on the sensitized endometrium, either the vascular changes or the decidual cell reaction or both. These questions and many others may be very relevant to our understanding of menstruation. Menstruation is a very unusual and unique phenomenon and there have been many theories regarding its causation. The present experiments do not provide any new ideas regarding its causation but by showing how a similar condition can be induced in rodents may allow experimental work to be done, which is not possible in women and by no means easy in other primates.

A very interesting aspect of menstruation, which has been noted many times, is the similarity of the process to inflammation. The similarity of decidual tissue to granulation tissue was also noted many years ago. The similarity starts with

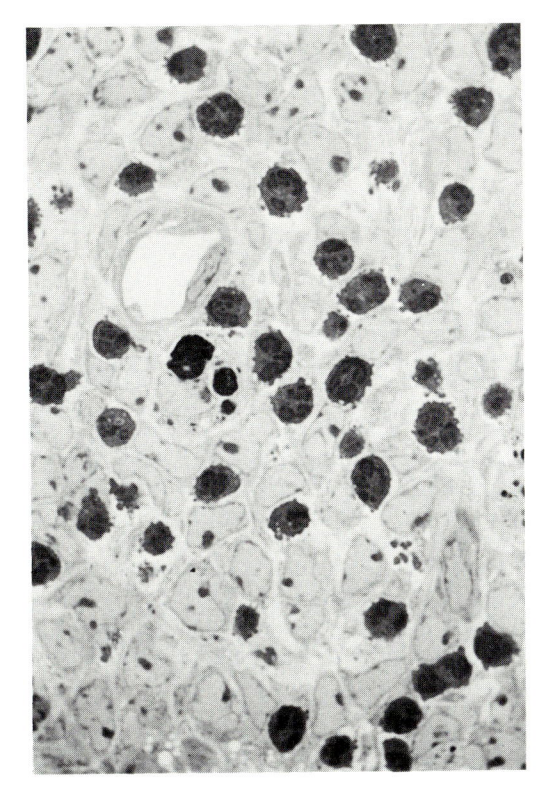

Figure 9. Section taken from the outer area of stroma shown in Figure 8, demonstrating massive migration of leucocytes. × 600

the vascular and cellular changes which prepare the uterus for implantation. Increased vascular permeability and oedema are well-recognized initial stages of inflammation. These changes in the rodent then lead in the presence of a blastocyst to the decidual reaction. However, if this is curtailed by stopping progesterone then migration of leucocytes and extravasation of blood cells occurs, which again are well-known components of inflammation.

It is tempting to speculate that the implantation reaction to the blastocyst has evolved from the inflammation response to a foreign body, although it must be remembered that the implantation reaction only occurs under very precise hormone conditions. If this were so then the human (and some other primates) must have evolved so that the reaction occurs not just in response to a blastocyst but in preparation for it. There are several examples of such preparatory physiological mechanisms; for example, the heart rate increases in anticipation of exercise and gut hormones cause release of insulin and calcitonin in preparation for increased glucose and calcium loads after a meal. The advantages of the latter are fairly obvious. The advantage to the human of anticipating the presence of a blastocyst with the consequent necessity of an inflammatory reaction in the uterus is less obvious.

References

Bryce, T.H., and Teacher, J.H. (1908). 'Contributions to the study of the early development and imbedding of the human ovum.' Glasgow.

Collawn, S.S., Rankin, J.C., Ledford, B.E., and Baggett, B. (1981). 'Ornithine decarboxylase activity in the artificially stimulated decidual cell reaction in the mouse uterus.' *Biol. Reprod.*, **24**, 528–533.

Dubransky, V., and Pohlman, G. (1961). 'Die Ultrastruktur des Korpus-Luteums und des Endometriums während das Cyclus.' *Arch. Gynaekol.*, **196**, 180–199.

Eichner, E., Goler, G.G., Reed, J., and Gordon, M.B. (1951). 'The experimental production and prolonged maintenance of decidua in the non-pregnant woman.' *Am. J. Obstet. Gynecol.*, **61**, 253–264.

Finn, C.A., and Bredl, J.C.S. (1973). 'Studies on the development of the implantation reaction in the mouse uterus: Influence of actinomycin D.' *J. Reprod. Fertil.*, **34**, 247–253.

Finn, C.A., and Hinchliffe, J.R. (1964). 'Reaction of the mouse uterus during implantation and deciduoma formation as demonstrated by changes in the distribution of alkaline phosphatase.' *J. Reprod. Fertil.*, **8**, 331–338.

Finn, C.A., and Keen, P.M. (1962). 'Studies on deciduomata formation in the rat.' *J. Reprod. Fertil.*, **4**, 215–216.

Finn, C.A., and Lawn, A.M. (1967). 'Specialized junctions between decidual cells in the uterus of the pregnant mouse.' *J. Ultrastruct. Res.*, **20**, 321–327.

Finn, C.A., and Martin, L. (1970). 'The role of the oestrogen secreted before oestrus in the preparation of the uterus for implantation in the mouse.' *J. Endocrinol.*, **47**, 431–438.

Finn, C.A., and Martin, L. (1972). 'Endocrine control of the timing of endometrial sensitivity to a decidual stimulus.' *Biol. Reprod.*, **7**, 82–86.

Finn, C.A., and Martin, L. (1973). 'Endocrine control of gland proliferation in the mouse uterus.' *Biol. Reprod.*, **8**, 585–588.

Finn, C.A., and Martin, L. (1976). 'Hormonal control of the secretion of the endometrial glands in the mouse.' *J. Endocrinol.*, **71**, 273–274.

Finn, C.A., and Pollard, R.M. (1973). 'The influence of the oestrogen secreted before oestrus on the timing of endometrial sensitivity and insensitivity.' *J. Endocrinol.*, **56**, 619–620.

Finn, C.A., and Pope, M. (1984). 'Vascular and cellular changes in the decidualized endometrium of the ovariectomized mouse following cessation of hormone treatment: a possible model for menstruation.' *J. Endocrinol.*, **100**, 295–300.

Gompel, C. (1964). 'Structur fine des mitochondries de la cellule glandulaire endometriale homaine au cours du cycle menstruel.' *J. Microsc.*, **3**, 427–436.

Green, C.A.L. (1979). 'Early pregnancy in guinea pigs.' *Ph.D. Thesis*, University of Liverpool.

Grosser, O. (1910). 'The development of the egg membranes and the placenta; menstruation.' In *Manual of Human Embryology* (eds. J. Keibel and P. Mall). W. Saunders, Philadelphia and London.

Hamperl, H. (1955). 'The granular endometrial stroma cells. A new cell type.' *J. Pathol. Bacteriol.*, **69**, 358–359.

Hunter, W. (1794). *An Anatomical Description of the Human Gravid Uterus and its Contents.* J. Johnson, London.

Kirby, D.R.S. (1960). 'Development of mouse eggs beneath the kidney capsule.' *Nature*, **187**, 707–708.

Krehbiel, R.H. (1937). 'Cytological studies of the decidual reaction in the rat during early pregnancy and in the production of deciduomata.' *Physiol. Zool.*, **10**, 212–238.

Lawn, A.M., Wilson, E.W., and Finn, C.A. (1971). 'The ultrastructure of human decidual and predecidual cells.' *J. Reprod. Fertil.*, **26**, 85–90.

Lejeune, B., Van Hoeck, J., and Leroy, F. (1981). 'Transmitter role of the luminal epithelium in the induction of decidualization in rats.' *J. Reprod. Fertil.*, **61**, 235–240.

Marshall, F.H.A. (1910). *The Physiology of Reproduction.* Longmans Green & Co., London, New York, Bombay & Calcutta.

Martin, L., and Finn, C.A. (1971). 'Oestrogen–gestagen interactions on mitosis in target tissues.' In *Basic Action of Sex Steroids on Target Organs* (eds. P.O. Hubinont, F. LeRoy and P. Galand). Karger, Basel.

Nelson, W.O., and Pfiffner, J.J. (1930). 'Experimental production of deciduomata in the rat by an extract of the corpus luteum.' *Proc. Soc. Exp. Biol. Med.*, **27**, 863–866.

Noyes, R.W., Hertig, A.T., and Rock, J. (1950). 'Dating the endometrial biopsy.' *Fertil. Steril.*, **1**, 3–25.

Pollard, R.M., and Finn, C.A. (1972). 'Ultrastructure of the uterine epithelium during the hormonal induction of sensitivity and insensitivity to a decidual stimulus in the mouse.' *J. Endocrinol.*, **55**, 293–298.

Psychoyos, A. (1960). 'La réaction déciduale est précedée de modifications précoces de la perméabilité capillaire de l'uterus.' *C.R. Seances Soc. Biol.*, **154**, 1384.

Sachs, L., and Shelesnyak, M.C. (1955). 'The development and suppression of polyploidy in the developing and suppressed deciduoma in the rat.' *J. Endocrinol.*, **12**, 146–155.

Seyle, H., and McKeown, T. (1935). 'Studies on the physiology of the maternal placenta of the rat.' *Proc. R. Soc. London Ser. B.*, **119**, 1–31.

Wilson, E.W. (1969). 'Alkaline phosphatase in pre-decidual cells of the human endometrium.' *J. Reprod. Fertil.*, **19**, 567–568.

The Luteal Phase
Edited by S.L. Jeffcoate
© 1985 John Wiley & Sons Ltd.

CHAPTER 4

Interaction between the embryo and the corpus luteum

D.K. EDMONDS
*Chelsea Hospital for Women,
Dovehouse Street,
London SW3 6LT, UK*

Introduction

Much speculation is enjoyed by reproductive physiologists regarding the means by which the embryo signals its presence to the mother, allowing recognition by the mother of the impending pregnancy. The importance of this process and the involvement of the corpus luteum has been the subject of intense research in the last 10 years, and the advent of techniques which have allowed evaluation of hormones at low concentrations has opened the door to increasing this understanding. But as our knowledge increases, so the number of questions to be answered increases, and the mechanism underlying the co-existence of embryo and mother becomes more fascinating.

Evidence suggests that the corpus luteum is essential in the maintenance of pregnancy at an early stage and this interaction is the subject of this chapter. It is also the intention to demonstrate that the reported comparatively poor conception rates in human are an underestimate and that these compare well with other mammalian species.

1. Corpus Luteum in Early Pregnancy

1.1. Indirect Evidence for the Necessity of the Corpus Luteum

The major evidence for the necessity of a normally functioning corpus luteum in the establishment and maintenance of early pregnancy comes from the association between infertility and the deficient luteal phase (see also Chapter 7). The sequential changes in endometrial morphology during the luteal phase of the ovulatory cycle are well-established, and if these morphological changes are retarded by 2 days or more then a diagnosis of a deficient luteal phase can be made which is associated with infertility (Jones, 1976). It is assumed that this physiological delay in endometrial maturation impedes implantation. That this defect has an endocrinological basis was demonstrated by studies examining progesterone levels in women with and without morphological evidence of a deficient luteal phase. The data were conclusive in documenting the association between inadequate progesterone production by the corpus luteum and endometrial retardation (Jones et al., 1974).

The production rate of progesterone, as determined by the area under the curve of progesterone levels in the luteal phase, is lower than that of fertile women, and some women who have recurrent abortions may have a similar pattern (Horta et al., 1977). Substitution therapy with progesterone in the luteal phase has resulted in successful pregnancies in these women (Jones, 1976).

1.2. Direct Evidence for the Necessity of the Corpus Luteum

The fundamental work of Csapo on the role of the corpus luteum in the establishment and maintenance of pregnancy is well known. The work was based on the results of luteectomy on established pregnancies on human subjects at various intervals following conception. The initial study (Csapo et al., 1972) measured progesterone production and pregnancy outcome following luteectomy between 42 and 74 days, and following luteectomy before 52 days gestation there was a rapid fall in progesterone and a resulting spontaneous abortion. In women failing to abort after 7 weeks' gestation, the levels of progesterone fell less markedly and subsequently rose again, thus indicating that the corpus luteum is essential for the maintenance of early pregnancy up to 7 weeks' gestation, but is dispensable thereafter. In a subsequent study (Csapo et al., 1973), luteectomized patients who were treated with 200 mg of progesterone by intramuscular injection in order to replace that produced by the corpus luteum, failed to abort, adding further evidence for the interdependence of the corpus luteum and conceptus.

2. Corpus Luteum Function in Conceptional Cycles

2.1. The Ovulatory Cycle Compared with the Conceptional Cycle

The steroid changes in the spontaneously ovulatory cycle and the interaction between gonadotrophin-releasing factors, gonadotrophin release and ovarian steroidogenesis have been reviewed (see Chapter 1). The follicular changes are dependent on adequate follicular stimulation by gonadotrophins, and subsequent ovulation under the influence of luteinizing hormone (LH) produces a normal corpus luteum. The subsequent fate of the corpus luteum is discussed elsewhere in this book (Chapter 2), but luteolysis eventually occurs and the subsequent morphological and endocrinological demise of the corpus luteum is inevitable.

However, in the conceptional cycle, the changes are somewhat different. The levels of progesterone and 17α-hydroxyprogesterone do not fall in either women (Yoshimi et al., 1969; Tulchinsky and Hobel, 1973; Corker et al., 1976) or monkeys (Knobil, 1973). The rise in progesterone production is significantly higher in the conceptual cycle by the 7th day post ovulation and has been reported as early as the 4th postovulatory day (Lenton et al., 1982), though this is disputed (see Chapter 7).

Oestrone and oestradiol also progressively increase from the luteal phase of the conceptional cycle until the 5th week of pregnancy, but thereafter there is a decrease in levels. Progesterone levels plateau at about the same time, but 17α-hydroxyprogesterone begins to decline. The steroid concentration rises again from between the 7th and 8th week of pregnancy as the placenta takes over production.

Thus, there is endocrinological evidence of an interaction between the presence of the conceptus and the corpus luteum in the conceptional cycle.

2.2. Embryonic Signals in Non-Primates

There is great variation between species of the endocrinological environment which governs maternal recognition of pregnancy. In domestic animals, the corpus luteum regresses after 16–20 days as a result of a luteolysin (probably prostaglandin (PG) $F_{2\alpha}$) produced by the uterus (Flint and Hillier, 1975). The presence of a viable embryo prevents this release, thereby allowing the corpus luteum to support the pregnancy in the early stages. There seems to be maternal recognition of pregnancy in domestic animals prior to implantation and this implies the production of a substance by the trophoblast which is luteotrophic. This may act directly on the uterus to prevent $PGF_{2\alpha}$ release (Peterson et al., 1976), or there may be a direct effect on the ovary (Mapletoft et al., 1976). The identification of the embryonic marker in the ewe remains unsolved, although it seems to be proteinaceous in nature.

In the pig, by contrast, oestrogen has been considered the luteotrophic substance, preventing $PGF_{2\alpha}$ release from the uterus. The secretion of oestrogen by the pig blastocyst begins between days 10 and 12 post coitus (Gadsby and Heap, 1978) and probably directly inhibits endometrial secretion of $PGF_{2\alpha}$. The oestrogen secreted, mainly oestrone sulphate, enters the circulation and acts on the hypothalamus and pituitary to increase LH release which is essential to prolong the life of the corpus luteum (Flint et al., 1979). Progesterone production is increased by this action (Watson and Maule-Walker, 1978).

In rabbits, progesterone production by the corpus luteum has been shown to be higher in pregnant does than in non-pregnant does in the luteal phase (Fuchs and Beling, 1974; Singh and Adams, 1978). In this species, chorionic gonadotrophin (CG) as well as oestrogen may be contributing to the luteotrophic effect (Haour and Saxena, 1974).

2.3. Embryonic Signals in Primates

The evidence for embryonic signals in primates is much less strong. There is good evidence, however, that rhesus monkey and human embyros produce CG, a protein capable of maintaining the life of the corpus luteum (Knobil, 1973; Ross, 1979). It seems that in the human hCG is produced by the blastocyst, certainly from the time of implantation, as demonstrated by elevated levels of progesterone by 7 days after the LH surge in conceptional as opposed to non-conceptional cycles (Corker et al., 1976), although Lenton et al. (1982) claim a significant rise as early as 2 days after the LH surge.

There is histological evidence that sinuses develop in the syncytiotrophoblast between 7 and 8 days following the LH surge and this may well be important for the transport of hCG to the maternal circulation. Whether hCG is secreted by the blastocyst prior to implantation is as yet unknown, but recent reports suggest evidence of another substance, early pregnancy factor, which may be involved in early recognition of pregnancy in women (Morton et al., 1977; Rolfe, 1982). This substance is yet to be isolated, but may well be produced soon after fertilization, prior to production of hCG.

The presence of hCG prevents corpus luteum regression, and this can be demonstrated by the administration of hCG to non-pregnant women. Immunization of primates against hCG causes failure of the establishment of pregnancy (Hearn et al., 1976), thereby elucidating the importance of hCG in the maintenance of early pregnancy. The rate of production of hCG rises during early pregnancy at a rapid rate to reach a maximum between 7 and 9 weeks of pregnancy. However, corpus luteum function begins to decline after the 6th week of pregnancy, probably due to 'down regulation' of the cells (Catt and Dufau, 1977).

3. Human Chorionic Gonadotrophin (hCG)

3.1. Chemistry

hCG is a glycoprotein containing 30% carbohydrate with a molecular weight of 36 700. It consists of two subunits: an α-subunit which is hormone non-specific, and a β-subunit which is specific; their molecular weights are 14 500 and 22 500, respectively. The α-subunit contains 92 amino acids, and the β-subunit 145. Some 80% of the β-subunit structure is similar to LH, but the terminal 30 amino acids are specific to hCG.

3.2. Biological Properties

Because of their structural similarities, LH and hCG are similar in their activities. The terminal 30 amino acid sequence is probably independent of the biological property of the molecule, though it appears to prolong the half-life of the peptide in the circulation.

Evidence suggests that the LH-, FSH- and TSH-like activity of hCG is due to structural similarities. The molecule may well be responsible for thyrotoxicosis in late pregnancy and choriocarcinoma.

3.3. Immunological Properties

hCG and both of its subunits are highly antigenic, and the similarity between all the glycoproteins has led to specific antisera being raised to enable the immunological measurement of hCG in the presence of other glycoproteins and metabolites of hCG. The major antigenic sites of the β-subunit of hCG are in the region of residues 21–23 (Swaminathan and Braunstein, 1978), and the basis of production of specific antisera to hCG has been in identifying the differences at this site between hCG and LH.

3.4. Measurement of hCG

There are four different techniques for measuring hCG: *in vivo* bioassays, immunological slide tests, radioimmunoassay and radioreceptor assay.

3.4.1. Biological tests.
In these methods suitable extracts of urine or serum are administered to appropriate laboratory animals and the ovarian response — hypertrophy, hyperaemia, and/or haemorrhage (Ascheim and Zondek, 1927) — is measured. The assays are cumbersome and time-consuming and lack the sensitivity of other methods.

3.4.2. Immunological slide tests.
The application of these tests followed the development of anti-hCG serum in rabbits. Urine samples are mixed with diluted antiserum and a suitable marker such as sensitized red cells, thus allowing the test to be carried out without the need of a laboratory animal. The sensitivity was not better than the biological tests, but the method was cheaper and simpler. The modern 'do-it-yourself' tests are a refinement of this technique but there is still a significant false positive and false negative rate.

3.4.3. Radioimmunoassay.
Many hCG radioimmunoasays are non-specific in that the antisera will cross-react with LH. Use of an antiserum raised against the β-subunit of hCG confers specificity, and levels of hCG as low as 5–6 mIU/Hml can be measured with both accuracy and precision. This technique is

widely used in quantitative measurement of hCG to assist in the diagnosis of early pregnancy, ectopic pregnancy, trophoblastic disease, etc.

3.4.4. Radioreceptor assay. Use of a specific receptor to a hormone as a binding reagent can give an estimate of the biologically active molecules. It does, however, cross-react with LH, but it is sensitive at levels of between 5 and 10 mIU/mlhCG.

3.5. hCG and its Receptor Site

hCG and LH are hydrophobic molecules which are located in the basal region of the lipid layer of cell membranes of gonadal cells (Bretscher and Raff, 1975). The majority of receptors occur in ovarian and testicular tissue and hCG and LH bind to the same receptor site. The presence of the glycoprotein on the receptor site induces the production of cyclic AMP, and this in turn stimulates the process within the cell to produce the steroid hormone.

The receptor sites appear in the ovary by about day 5 of the cycle and increase in number until they reach a maximum at day 17 of the cycle and their numbers remain unchanged until the end of the cycle. The affinity of the receptor binding of LH or hCG seems to remain constant throughout the cycle (Rajaniemi *et al.*, 1981).

3.6. Biological Role of hCG in Pregnancy

hCG appears to act in several different ways during pregnancy. Hirono *et al.* (1972) showed that hCG may have a direct inhibitory role on the median eminence of the hypothalamus, reducing the release of FSH and LH during pregnancy and thus preventing ovulation. It may also have an influence on the intrauterine immuno privilege which the embryo enjoys during implantation and foetal life. This may be via an effect on maternal lymphocytes and their ability to produce antibodies, and there is some evidence to suggest a synergistic role with human placental lactogen (Kitzmiller and Rocklin, 1980; Porter and Amoroso, 1977). There is good evidence for a role in foetal gonad development (Clements *et al.*, 1976), but its effect on the foetal adrenal to produce dehydroepiandrosterone remains tentative (Chattergee and Munro, 1977). The classical role, however, relates to its direct action as a luteotrophin, stimulating the production of increasing amounts of progesterone from the ovary during early pregnancy; whether it has any action on placental steroidogenesis is debatable.

4. Early Embryonic Mortality

4.1. Mammalian Evidence

Much information has been recorded about embryonic loss in farm animals, e.g. pigs, sheep and cattle. Most of the data are derived from investigations at various stages of gestation and comparing the number of corpora lutea with the number of embryos present. In the pig, evidence suggests that as many as 40% of fertilized oocytes will have suffered embryonic mortality; in the sheep, 25–30% of embryos fail to develop into term lambs, and in cattle, the embryonic loss rate is about 20% with a further 6% being lost from day 34 onwards (Boyd, 1965).

The aetiology of this conceptual loss remains unknown although it would seem that genetic factors may be significant.

4.2. Human Evidence

The collection of data on women with regard to early embryonic mortality is difficult, and the majority of the information on fecundability is derived from epidemiological evidence.

Using church records from the 18th century, Henry (1965) calculated that the chance of producing a full-term infant per menstrual cycle (fecundability) was 23%. Another French study, from Vincent (1961), using data from the early 20th century, showed a similar figure of 27%. MacLeod and Gold (1953) followed the attempts of 428 women in New York to conceive and they found that 83% were pregnant in 6 months with a fecundability of 25%. Recent work by Vessey et al. (1976) has added still further evidence that 21% of women will deliver a term child 11 months after ceasing barrier methods of contraception. So, the evidence over the last 200 years remains remarkably consistent with regard to fecundability.

It would seem, therefore, that such a poor fecundability in humans, when compared to other mammals, would imply a considerable embryonic loss. Roberts and Lowe (1975) used a mathematical estimation to suggest that 78% of fertilizations failed to result in a live birth. The definitive histological work in this field, however, came from Hertig et al. (1952); their examination of hysterectomized uteri indicated an embryonic loss rate of 40% post implantation.

Using hCG as a marker of trophoblastic activity, Edmonds et al. (1982) studied 198 ovulatory cycles from a normal population of women who were attempting conception. Fecundability in this population was 22% but the risk of pregnancy in exposed cycles was 59.6%. Over 60% of these conceptuses will be lost prior to the 12th week of pregnancy, however, and the majority (91.7%) will be lost subclinically without the knowledge of the mother. This method of detecting trophoblastic activity is limited to the time of implantation, but work from Rolfe (1982) using early pregnancy factor suggested that fertilization can be detected in humans in some 69% of ovulatory cycles.

One may speculate that these large prenatal losses are due to errors in gametogenesis with non-disjunction leading to trisomy or monosomy, and when these chromosomal errors are related to gestation the incidence of abnormalities falls with increasing gestation during the first trimester. Extrapolating the data backwards, one would expect 90% of post-implantation embryonic losses prior to menstruation to have this type of abnormality. Other causes include defects at fertilization with abnormalities like triploidy being a major cause of loss prior to implantation; abnormality in the phenotypic development of the embryo, although important in later causes of embryonic loss, probably accounts for less at the earlier stages.

5. Conclusions

There is considerable direct and indirect evidence that maternal recognition of the presence of a fertilized oocyte is vital to the successful course of embryonic development and a successful pregnancy. This recognition is complex but partially explicable by the effects of chorionic gonadotrophin. The interaction between the embryo and the corpus luteum, via this hormonal mediator, is fundamental in this recognition process as changes in the corpus luteum and its function during a conceptional cycle allow the pregnancy to continue.

However, the recognition process may be far from perfect, as evidenced by the high embryonic loss in animals, but especially in humans. Or is it so imperfect? The protective mechanism and the changes in the immunological status of the mother would imply that if gametogenesis is abnormal, the recognition can be overcome and the pregnancy rejected. When comparison is made between the number of chromosomally abnormal children born, and the number of chromosomally abnormal spontaneous abortions, the process of recognition and conceptus selection can be seen to be much more complex than merely the presence of hCG.

Our understanding of the interaction between the embryo and the mother is still in its infancy, but our knowledge is slowly growing and presents a future of fascination.

References

Ascheim, S., and Zondek, B. (1927). *Klin. Wochenschr.*, **6**, 1322.
Boyd, H. (1965). 'Embryonic death in cattle, sheep and pigs.' *Vet. Bull.*, **35**, 251–266.
Bretscher, M.S., and Raff, M.C. (1975). 'Mammalian plasma membranes.' *Nature*, **258**, 43–45.
Catt, K.J., and Dufau, M.L. (1977). 'Peptide hormone receptors.' *Annu. Rev. Physiol.*, **39**, 529–557.

Chattergee, M., and Munro, H.N. (1977). 'Structure and biosynthesis of human placental peptide hormones.' *Vitam. Horm.*, **35**, 149–208.

Clements, J.A., Reyes, F.I., Winter, J.S.D., and Faiman, C. (1976). 'Studies on human sexual development. III. Fetal pituitary and serum and amniotic fluid concentrations of LH, CG and FSH.' *J. Clin. Endocrinol. Metab.*, **42**, 9–19.

Corker, C.S., Michie, E., Hobson, B., and Parboosingh, J. (1976). 'Hormonal patterns in conceptual cycles and early pregnancy.' *Br. J. Obstet. Gynaecol.*, **83**, 489–494.

Csapo, A.I., Pulkinnon, M.O., Rutter, B., Sauvage, J.P., and Wiest, W.G. (1972). 'The significance of the human corpus luteum in pregnancy maintenance.' *Am. J. Obstet. Gynecol.*, **112**, 1061–1067.

Csapo, A.I., Pulkinnon, M.O., and Wiest, W.G. (1973). 'Effects of luteectomy and progesterone replacement therapy in early pregnant patients.' *Am. J. Obstet. Gynecol.*, **115**, 759–765.

Edmonds, D.K., Lindsay, K.S., Miller, J.F., Williamson, E., and Wood, P.J. (1982). 'Early embryonic mortality in women.' *Fertil. Steril.*, **38**, 447–453.

Flint, A.P.F., and Hillier, K. (1975). In *Prostaglandins and Reproduction* (ed. S.M.M. Karim) pp. 271–308. MTP Press, Lancaster.

Flint, A.P.F., Burton, R.D., Gadsby, J.E., Saunders, P.T.K., and Heap, R.B. (1979). 'Hormone secretion in early pregnancy.' In *Maternal Recognition of Pregnancy* (ed. J. Whelan), *Ciba Foundation Symposium*, **64**, 53–74. Excerpta Medica, Amsterdam.

Fuchs, A.R., and Beling, C. (1974). 'Evidence for early ovarian recognition of blastocysts in rabbits.' *Endocrinology*, **95**, 1054–1058.

Gadsby, J.E., and Heap, R.B. (1978). In *Novel Aspects of Reproductive Physiology* (eds. C.H. Spilman and J.W. Wilks), pp. 265–285. S.P. Medical and Scientific Books, New York.

Haour, F., and Saxena, B.B. (1974). 'Detection of a gonadotrophin in rabbit blastocyst before implantation.' *Science*, **185**, 444–445.

Hearn, J.P., Short, R.V., and Lunn, S.F. (1976). In *Physiological Effects of Immunity Against Reproductive Hormones* (eds. R.G. Edwards and M.H. Johnson), pp. 229–243. Cambridge University Press.

Henry, L. (1965). 'French statistical research in natural fertility.' In *Public Health and Population change* (eds. M.C. Sheps and J.C. Ridley), University of Pittsburg Press.

Hertig, A.T., and Rock, J. (1945). *Contrib. Embryol.*, **31**, 65–84.

Hertig, A.T., Rock, J., Adams, M.C., and Menkin, M.C. (1952). 'Thirty four fertilized ova, good, bad and indifferent recovered from 210 women of known fertility.' *Pediatrics*, **23**, 202–210.

Hirono, M., Igarashi, M., and Matsumoto, S. (1972). 'The direct effect of HCG upon pituitary gonodotrophin secretion.' *Endocrinology*, **90**, 1214–1219.

Horta, J.L., Fernandez, J.G., DeLeon, B.S., and Cortes-Gallegos, V.C. (1977). 'Direct evidence of luteal phase deficiency in women with habitual abortion.' *Obstet. Gynecol.*, **49**, 705–708.

Jones, G.S. (1976). 'The luteal phase defect.' *Fertil. Steril.*, **27**, 351–356.

Jones, G.S., Absel, S., and Wentz, A.C. (1974). 'Serum progesterone values in lutal phase defects.' *Obstet. Gynecol.*, **44**, 26–34.

Kitzmiller, J.L., and Rocklin, R.E. (1980). 'Lack of suppression of lymphocyte MIF production by estradiol, progesterone and human chorionic gonadotrophin.' *J. Reprod. Immun.*, **1**, 297–301.

Knobil, E. (1973). 'On the regulation of the primate corpus luteum.' *Biol. Reprod.*, **8**, 246–258.

Lenton, E.A., Sulaiman, R., Sobowale, O., and Cooke, I.D. (1982). 'The human menstrual cycle: plasma concentrations of prolactin, LH, FSH, oestradiol and progesterone in conceiving and non-conceiving women.' *J. Reprod. Fertil.*, **65**, 131–139.

MacLeod, J., and Gold, R.Z. (1953). 'The male factor in fertility and sterility.' *Fertil. Steril.*, **4**, 10–33.

Mapletoft, R.J., Lapin, D.R., and Ginther, O.J. (1976). 'The ovarian artery as the final component of the local luteotrophic pathway between a gravid uterine horn and ovary in ewes.' *Biol. Reprod.*, **15**, 414–421.

Morton, H., Rolfe, B.E., Clunie, G.J.A., Anderson, M.J., and Morrison, J. (1977). 'An early pregnancy factor detected in human serum by rosette inhibition test.' *Lancet*, **1**, 394.

Peterson, A.J., Tenit, H.R., Fairclough, R.J., Havik, P.G., and Smith, J.F. (1976). 'Jugular levels of 13, 14-keto-dihydro-15-keto-prostaglandin F and progesterone around luteolysis and early pregnancy in the ewe.' *Prostaglandins*, **12**, 551–558.

Porter, D.G., and Amoroso, E.C. (1977). 'The endocrine function of the placenta. In *Scientific Foundations of Obstetrics and Gynaecology* (eds: E. Phillips, J. Barnes and M. Newton), p. 675. Heinemann, London.

Rajaniemi, H.J., Ronnberg, L., Kauppila, A., Ylostalo, P., Jalkanen, M., Saastamoinen, J., Selander, K., Pystynen, P., and Vihko, R. (1981). 'LH receptors in human ovarian follicles and corpora lutea during menstrual cycle and pregnancy.' *J. Clin. Endocrinol. Metab.*, **52**, 307–311.

Roberts, C.J., and Lowe, D.B. (1975). 'Where have all the conceptions gone?' *Lancet*, **i**, 498.

Rolfe, B.E. (1982). 'Detection of fetal wastage.' *Fertil. Steril.*, **37**, 655–660.

Ross, G.T. (1979). 'The biochemistry of human chorionic gonadotrophin.' In *Maternal Recognition of Pregnancy* (ed. J. Whelan), *Ciba Foundation Symposium*, **64**, 191–201. Excerpta Medica, Amsterdam.

Singh, M.M., and Adams, C.E. (1978). 'Luteotrophic effect of the rabbit blastocyst.' *J. Reprod. Fertil.*, **53**, 331–333.

Swaminathan, N., and Braunstein, G.D. (1978). 'Location of major antigenic sites of the subunit of human chorionic gonadotrophin.' *Biochemistry*, **17**, 5832–5837.

Tulchinsky, D., and Hobel, C.J. (1973). 'Plasma human chorionic gonadotrophin, estrone, estradiol, progesterone and 17α-hydroxyprogesterone in human pregnancy. III. Early pregnancy.' *Am. J. Obstet. Gynecol.*, **117**, 884–893.

Vessey, M., Doll, R., Peto, R., Johnson, B., and Wiggins, P. (1976). 'Long-term follow-up study of women using different forms of contraception – an interim report.' *J. Biosoc. Sci.*, **8**, 373–427.

Vincent, P. (1961). *Recherches sur la Fecondité Biologique*. Press Universitaires de France, Paris.

Watson, J., and Maule-Walker, F.M. (1978). 'Progesterone production by the corpus luteum of the early pregnant pig during superfusion *in vitro* with PGF, LH and oestradiol.' *J. Reprod. Fertil.*, **52**, 209–212.

Yoshimi, T., Strott, C.A., Marshall, J.R., and Lipsett, M.B. (1969). 'Corpus luteal function in early pregnancy.' *J. Clin. Endocrinol. Metab.*, **29**, 225–230.

The Luteal Phase
Edited by S.L. Jeffcoate
© 1985 John Wiley & Sons Ltd.

CHAPTER 5

Prolactin and the corpus luteum

ALAN S. MCNEILLY
MRC Reproductive Biology Unit,
37 Chalmers Street,
Edinburgh EH3 9EW, UK

Introduction

Following the purification and the subsequent development of specific radio-immunoassays for human prolactin in the early 1970's, it was quickly recognized that up to 20% of cases of secondary amenorrhoea were asociated with plasma levels of prolactin in excess of those seen in women with normal menstrual cycles. Since normalization of these excessive levels of prolactin either by surgical removal of prolactin-secreting tumors or by pharmacological treatment with dopamine agonists led to resumption of normal ovarian function, it was assumed that the raised plasma levels of prolactin were responsible wholly or in part for the failure of normal ovarian activity. If such an action was to occur then it would be logical to expect prolactin also to play a role in normal ovarian function. However, this latter area has received remarkably little attention in the human, although there is irrefutable evidence that prolactin is an essential component of the complex of hormones necessary to maintain normal corpus luteum function in some other species. This review will briefly outline the evidence that prolactin is involved in the maintenance of corpus luteum function by presenting evidence gathered from species other than the human, and then examine the evidence that prolactin is involved in a similar manner in man.

1. Prolactin Receptors on the Corpus Luteum

It is clear from many studies in tissues in which prolactin exerts a defined function that these actions are relayed via a receptor present on the cell membrane. While there remain some unresolved technical difficulties in demonstrating binding of prolactin, several techniques including conventional radioreceptor techniques with ^{125}I-labelled prolactin and homogenates or cell membrane preparations of corpora lutea, localization of prolactin using topical autoradiography or the use of immunocytochemical methods with antiserum to prolactin on the prolactin-receptor have shown specific receptors for prolactin on the corpora lutea from the rat (Richards and Williams, 1976; Nolin, 1980; Dunaif et al., 1982), hamster (Oxberry and Greenwald, 1982), pig (Rolland et al., 1976), wallaby (Sernia and Tyndale-Biscoe, 1979), cow (Saito and Saxena, 1975), and human (McNeilly et al., 1980b). The affinity of the receptors are similar to those on the mammary gland, an accepted prolactin target tissue. In the pig (Rolland et al., 1976) binding to the corpus luteum increases during pregnancy, while in the rat (Chang, 1976; Richards and Williams, 1976) binding increases until day 9 of pregnancy and then declines steadily until parturition on day 21. In the rat the induction of prolactin receptors on the corpus luteum is dependent on luteinizing hormone (LH) and not on prolactin (Holt et al., 1976). In contrast, prolactin is responsible for the induction of LH receptors in the newly formed corpora lutea following ovulation (Holt et al., 1976) although this may involve a synergism with follicle-stimulating hormone (FSH) (Casper and Erickson, 1981). Specific binding of prolactin has been demonstrated in the human corpus luteum, but this was mainly confined to the early luteal phase (Figure 1) (McNeilly et al., 1980b). The reason for this remains unexplained but it has been confirmed using cell membrane preparations from human corpora lutea (T.A. Bramley, unpublished observations).

Thus the presence of specific prolactin receptors on the corpus luteum suggests that prolactin may be important for the maintenance of luteal function.

2. Changes in Blood Concentrations of Prolactin

In the majority of species so far investigated no specific changes — either increases or decreases — in the blood concentrations of prolactin have been identified. This is particularly so in the primates, including the chimpanzee (Reyes et al., 1975), rhesus monkey (Butler et al., 1975), marmoset (McNeilly et al., 1981) and gorilla (Nadler et al., 1979), and the majority of studies in women (Ehara et al., 1973; McNeilly and Chard, 1974; Lenton et al., 1982). In women there is evidence for an increase in prolactin levels around mid-cycle (Robyn et al., 1976; Franchimont et al., 1976) although this was not apparent in other studies (see McNeilly, 1980). More recently we have shown a clear increase in prolactin levels associated with the preovulatory surge of LH when frequent blood samples were taken (Djahanbakhch et al., 1984). The increase in prolactin was coincident with the LH surge and there was no apparent relationship between the changes in prolactin and plasma levels of oestradiol. It would appear that this increase in prolactin is related to an alteration, presumably a suppression, in hypothalamic dopamine turnover, necessary for the release of

PROLACTIN (PRL) BINDING TO CL
DURING LUTEAL PHASE IN RELATION
TO CL WEIGHT AND SERUM PRL

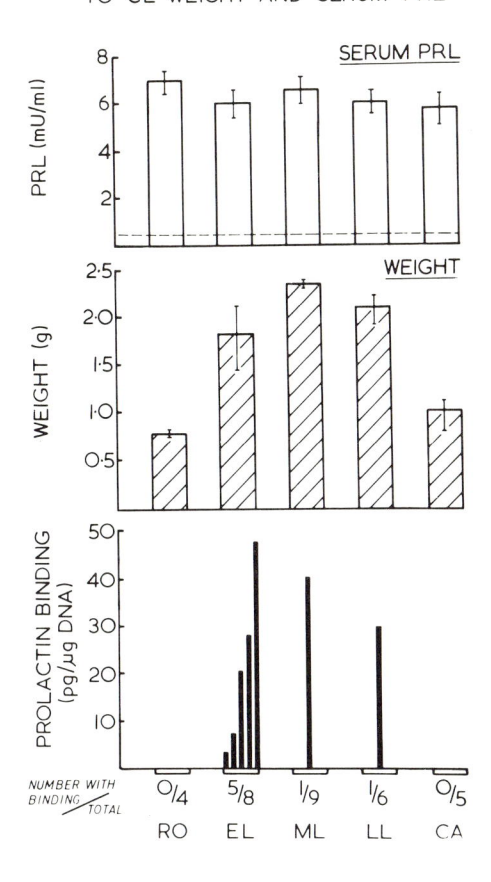

Figure 1. The binding of ^{125}I-labelled human prolactin to homogenates of human corpora lutea, corpus luteum weight (mean ± SEM) and serum levels of prolactin (mean ± SEM) collected at different stages of the luteal phase of the human menstrual cycle. The number of corpora lutea showing binding in relation to those examined are shown for each stage of the luteal phase. RO, recent ovulation; EL, early luteal phase; ML, mid-luteal phase; LL, late luteal phase; CA, corpus albicans. (From McNeilly et al., 1980)

luteinizing hormone releasing hormone (LHRH) at the time of the pre-ovulatory surge of LH (Judd et al., 1978; McNeilly, 1980).

A similar increase in prolactin around the time of the preovulatory surge of LH has been observed in several other species (see McNeilly, 1980; McNeilly, 1984) but only in rodents are specific changes in prolactin apparent in the luteal phase. The oestrous cycle of many rodents, including in particular the rat and mouse, is typified by the formation of corpora lutea of limited life span and ability to secrete progesterone (Rothchild, 1981). Only when prolactin levels increase in the luteal phase do these corpora lutea become fully functional developing the ability to secrete large amounts of progesterone and maintain pregnancy. In the rat, this increase in prolactin secretion occurs as the release of twice daily diurnal and nocturnal surges of prolactin, triggered by mating and continuing until day 11 of the 22-day pregnancy (Smith and Neill, 1976). Cervical stimulation in the absence of mating (Freeman *et al.*, 1974) or prolactin from an ectopic pituitary gland (de Greef and Zeilmaker, 1978) will also initiate these surges although progesterone is required for their maintenance (de Greef and Zeilmaker, 1978; Gorospe and Freeman, 1981). The luteotrophic effect of prolactin is first evident on the morning of day 3 of pregnancy or pseudopregnancy (Morishige and Rothchild, 1974; Day *et al.*, 1980) since suppression of prolactin by bromocriptine on days 1 and 2 has no effect while suppression on day 3 results in the termination of pregnancy (Smith *et al.*, 1976).

Treatment of rats with antiserum to LH also shows that the corpora lutea are prolactin-dependent but LH-independent until day 9 of pregnancy, while between days 9 and 12 both prolactin and LH are essential (Garris and Rothchild, 1980; Garris *et al.*, 1980). From day 12 until term, corpora lutea are maintained by placental hormones (Rothchild, 1981).

Similar, though less extensive data, suggest that, in addition to the rat, the corpora lutea of the mouse (Dominic, 1966), hamster (Greenwald, 1967) and vole (Milligan and McKinnon, 1976) are also totally dependent on prolactin for the induction and maintenance of progesterone secretion during early pregnancy.

3. Prolactin and Steroidogenesis in the Corpus Luteum

Previous chapters (see Chapters 1 and 2) have presented evidence that the synthesis and secretion of ovarian steroids is predominantly under the control of LH with FSH acting within the follicle to induce an active aromatase enzyme system in the granulosa cells of the follicle to allow conversion of androgens to oestrogens. The role of FSH in the corpus luteum remains unclear. Apart from rodents, where prolactin is essential for the maintenance of progesterone secretion by the corpus luteum, there is still considerable doubt as to whether prolactin is essential for or even involved in steroidogenesis. This may in part arise from the expectation of many authors that prolactin *per se* should stimulate active steroid secretion by the corpus luteum. Before discussing the effects of prolactin in other species where its role remains controversial, the mechanisms of action of prolactin within the rat corpus luteum — where it is essential — will first be examined.

3.1. Prolactin and the Rat Corpus Luteum

This section will only briefly summarize the essential effects of prolactin since it has been reviewed in detail previously (Rothchild, 1981). As described above (section 1), in the rat, prolactin, with FSH and oestradiol, induces and maintains the LH receptor on the corpus luteum and is essential for the induction and maintenance of progesterone secretion by the corpus luteum (section 2). At the time of the prolactin-induced increase in luteal cell LH receptor (Richards and Williams, 1976) there is an enhancement of luteal LH-dependent adenylate cyclase both of which develop in parallel with the capacity of the corpus luteum to secrete progesterone (Day *et al.*, 1980). In spite of the essential role of prolactin in corpus luteum function there is no evidence that prolactin alone can stimulate progesterone production *in vitro* from luteal cells isolated during early pregnancy (Wu *et al.*, 1976; Murakami *et al.*, 1982) although LH may cause an increase in progesterone secretion (Murakami *et al.*, 1982; Wu *et al.*, 1976).

Within the rat corpus luteum prolactin appears to maintain the level of cholesterol esters, the precursors for steroidogenesis, by maintaining cholesterol ester synthetase and cholesterol ester hydrolase (Armstrong *et al.*, 1970; Hashimoto and Wiest, 1969). Indeed, the nocturnal surge of prolactin in the rat is followed 6 h later by an increase within the corpus luteum of cholesterol ester hydrolase liberating cholesterol to serve as a precursor for progesterone production (Klenscke and Brinkley, 1980a, b). Prolactin also affects the amount of progesterone secreted by inhibiting the action of the enzyme 20α-hydroxysteroid dehydrogenase (Wiest *et al.*, 1968). This enzyme catalyses the reduction of progesterone to its inactive form 20α-dihydroprogesterone, although the importance of this modulation of progesterone secretion remains unclear since a simple direct reciprocal relationship between progesterone and 20α-dihydroprogesterone does not always exist, particularly at the end of pregnancy in the rat (Lacy *et al.*, 1976; Labhetswar and Watson, 1974).

In mid-pregnancy when the corpus luteum in the rat is dependent on both prolactin and LH, prolactin regulation of the LH receptor becomes distinct from the maintenance of progesterone secretion. Hypophysectomy and hysterectomy results in a decline in LH receptors within 48 h which can be prevented by administration of prolactin (Gibori and Richards, 1978). In contrast, progesterone secretion decreased within 24 h and was not maintained by prolactin. However, when oestradiol was given at the same time as prolactin, progesterone secretion was maintained. It is probable, therefore, that LH acts on the corpus luteum by stimulation of oestradiol or its immediate androgen precursor (Gibori *et al.*, 1979; Gibori and Richards, 1978) which synergizes with prolactin to maintain luteal progesterone production. This synergism appears to occur by prolactin maintaining the levels of the cytoplasmic receptor for oestradiol within the corpus luteum (Gibori *et al.*, 1979) although how prolactin achieves this remains unknown.

Thus, in the rat corpus luteum where prolactin is known to be essential for progesterone, it acts to increase LH receptor numbers, maintain steroid precursor pools of cholesterol ester, reduce the levels of 20α-hydroxysteroid dehydrogenase and maintain the intracellular levels of cytoplasmic oestradiol receptor. It acts in a synergistic manner but does not itself stimulate steroidogenesis directly.

3.2. In Vivo Studies in Species other than the Rat

As we have seen, suppression of plasma levels of prolactin in the rat leads to failure of corpus luteum function. In contrast, suppression of prolactin by bromocriptine or ergocornine in sheep (Louw et al., 1974; Niswender, 1974) and cows (Hoffman et al., 1974) or administration of prolactin during the luteal phase of the cycle in sheep (Karsch et al., 1971; McCracken et al., 1971), cow (Donaldson et al., 1965; Smith et al., 1957) and pig (Anderson, 1966) fails to affect progesterone secretion by the corpus luteum, suggesting that prolactin is not involved in luteal maintenance in this species.

However, following hypophysectomy in the sheep (Denamur et al., 1973), goat (H. Buttle, personal communication) and ferret (Murphy, 1979), maintenance of full luteal function and normal progesterone secretion is only achieved by the administration of both prolactin and LH. In the sheep, prolactin alone is able to maintain some luteal activity while LH alone is ineffective (Denamur et al., 1973) (Figure 2). Luteal function in sheep is also maintained after pituitary stalk section on day 3 of the luteal phase (Denamur et al., 1966), a situation in which plasma levels of LH are undetectable but prolactin levels are maintained or elevated (Kann and Denamur, 1974).

In the rhesus monkey, suppression of prolactin secretion by bromocriptine reduces luteal phase progesterone secretion and abolishes the increase in oestradiol, suggesting an important role for prolactin (Espinosa-Campos et al., 1975). The amount of prolactin required is probably small, however, since near normal luteal function can be maintained after hypophysectomy of rhesus monkeys in the early luteal phase where plasma levels of both LH and prolactin become undetectable (Asch et al., 1982). However, the detection limit of the prolactin assay in this study was 3 ng/ml, a level well above that necessary for luteal maintenance in the ewe. Prolactin appears to be necessary for the maintenance of progesterone secretion by the corpus luteum of pregnancy which is maintained post partum in the lactating rhesus monkey (Weiss et al., 1976).

The situation in women is unclear since bromocriptine-induced suppression of prolactin levels in normoprolactinaemic normally cycling women has been shown either to have no effect (Del Pozo et al., 1975) or to lead to corpus luteum insufficiency (Schulz et al., 1980). However, these differences may relate not to the effect on prolactin but to a varying direct effect of bromocrip-

CL FUNCTION IN HYSTERECTOMIZED EWES

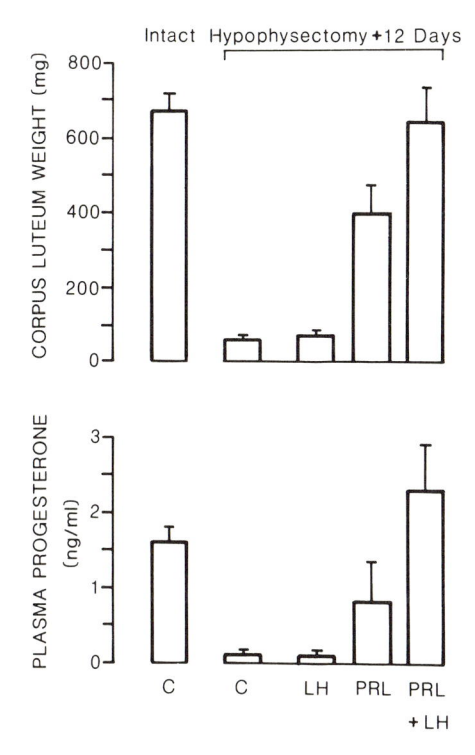

Figure 2. Corpus luteum weight and plasma levels of prolactin in intact (c) ewes and in ewes hypophysectomized and untreated (c) or treated for 12 days with LH, prolactin (PRL) or PRL + LH. (Redrawn from Denamur *et al.*, 1973)

tine on gonadotrophin secretion. Certainly the amount of prolactin needed to maintain normal corpus luteum function is very small since normal follicle growth ovulation and normal corpus luteum function can be induced in hypophysectomized women (Gemzell, 1975) even when plasma levels of prolactin are well below (< 15 mU/l), the lower limit seen in normal menstruating women (Figure 3).

3.3. Prolactin and the Primate Corpus Luteum: Studies In Vitro

This area has received surprisingly little attention. *In vitro* studies with luteal cells isolated from corpora lutea of the macaque suggested that prolactin may synergize with LH in the maintenance of progesterone secretion by the corpus luteum (Stouffer *et al.*, 1980).

In the human, studies with granulosa cells that have undergone luteinization

INDUCTION OF OVULATION AND PREGNANCY IN A HYPOPROLACTINAEMIC WOMAN
(PRL <15mU/l)

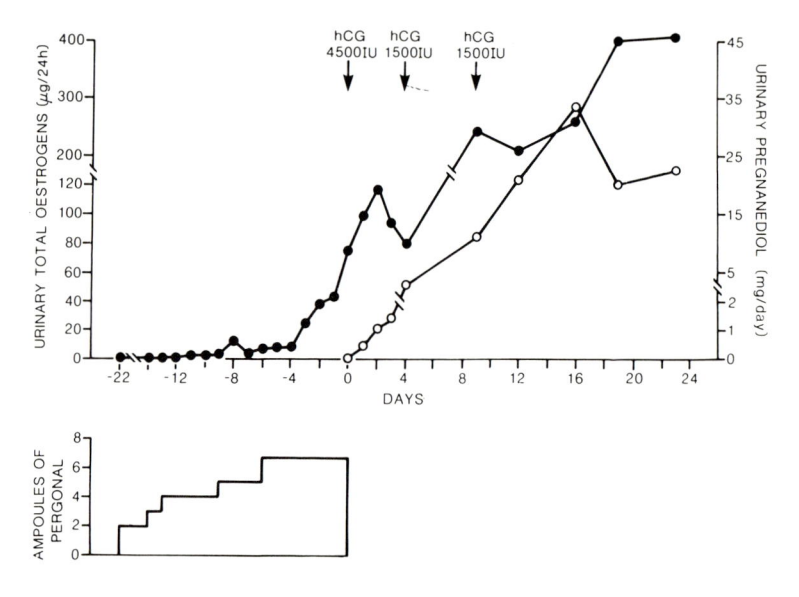

Figure 3. Changes in urinary levels of total oestrogens (●) and pregnanediol (○) during the induction of ovulation with Pergonal and hCG in a hypophysectomized women with hypoprolactinaemia (plasma level of prolactin < 15 mU/l). The resulting pregnancy was uneventful with plasma levels of prolactin rising only to 80 mU/l at term (normal range at term 1500–9000 mU/l). (A.S. McNeilly and D.T. Baird, unpublished observations).

in vitro and are secreting maximal amounts of progesterone suggest that prolactin is essential for the maintenance of progesterone secretion (McNatty *et al.*, 1974). However, it is debatable whether these granulosa cells luteinized *in vitro* are sufficiently similar to granulosa lutein cells within the corpus luteum for firm conclusions to be drawn from these studies. *In vitro* studies using minced luteal tissue suggests that, while prolactin does not affect either progesterone or oestradiol secretion, it does enhance chorionic gonadotrophin (hCG) stimulated steroid production (Hunter, 1980) (Figure 4A). In contrast, studies with isolated luteal cells *in vitro*, while confirming the absence of effect of prolactin alone, failed to demonstrate any synergistic effect of prolactin on hCG-stimulated steroidogenesis (Tann and Biggs, 1983) (Figure 4B). However, in the latter study the dose of hCG used (10 IU/ml) was well above that which would induce a maximal steroid response (Richardson and Masson, 1981) and any synergism with prolactin may have been masked. Clearly more studies are required to resolve these differences.

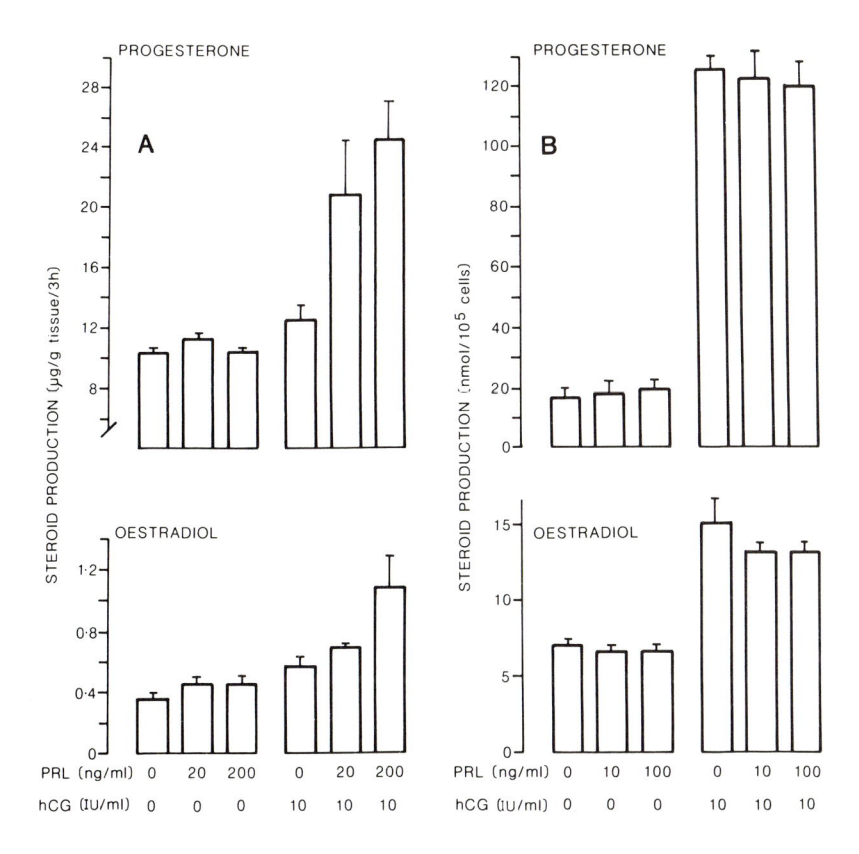

Figure 4. Effect of prolactin and hCG alone and in combination on the secretion of progesterone and oestradiol from human corpora lutea maintained *in vitro*. A. Results from Hunter (1980) using minces of luteal tissue. B. (Redrawn) from Tann and Biggs (1983) using collagenase-dispersed luteal cells

4. Corpus Luteum Function in Hyperprolactinaemic States

4.1. Pathological Hyperprolactinaemia

The demonstration that up to 20% of women presenting with secondary amenorrhoea were hyperprolactinaemic suggested that abnormal increases in prolactin may be directly involved in suppression of fertility (Thorner, 1977). While the aetiology of hyperprolactinaemia remains unclear it has been observed that, as the disease progresses and prolactin levels increase, corpus luteum function declines with an increase in the incidence of menstrual cycles with inadequate corpus luteum function before eventually becoming amenorrhoeic

(Reyes *et al.*, 1977; Thorner, 1977). Thus it is possible that there is a direct effect of high levels of prolactin on corpus luteum function. However, several lines of evidence suggest that this is not the case. In women in whom luteal phase defects have been shown to occur, there was no consistent correlation with hyperprolactinaemia which could explain the inadequate luteal function (Sarris *et al.*, 1978; Vanrell and Balasch, 1983), although associations have been found (e.g. Seppala *et al.*, 1976). Simiarly, when hyperprolactinaemia is induced during the luteal phase of the menstrual cycle itself either by using dopamine receptor blocking drugs such as sulpiride (Robyn *et al.*, 1977) or by withdrawing bromocriptine treatment in hyperprolactinaemic women (Bennink, 1979), corpus luteum function is unaffected.

Studies with human granulosa cells which were luteinized *in vitro* had suggested that high levels of prolactin could reduce progesterone production (McNatty *et al.*, 1974). However, more recent studies have not confirmed these initial observations (Edwards *et al.*, 1982; Swanston *et al.*, 1984), although in the latter study granulosa cells were harvested from preovulatory follicles 30 h after the onset of the preovulatory LH surge and luteinization had been initiated *in vivo* (Swanston *et al.*, 1984).

Thus it would appear that, once the corpus luteum is formed, high levels of prolactin do not affect luteal function adversely. The formation of inadequate corpora lutea in hyperprolactinaemic states is most probably a consequence of disturbance of follicular development in which prolactin may play some part (McNeilly *et al.*, 1982a).

4.2. *Lactational Infertility*

There is abundant evidence that lactation suppresses fertility by delaying the resumption of ovarian activity in many species (see Lamming, 1978; McNeilly, 1979, 1980). In man, lactational infertility is associated with hyperprolactinaemia (Delvoye *et al.*, 1978; McNeilly *et al.*, 1980a), and studies in Africa have shown a correlation between percentage of women with hyperprolactinaemia and those with amenorrhoea (Delvoye *et al.*, 1978; Duchen and McNeilly, 1980). There is no doubt that both the degree of hyperprolactinaemia and the duration of amenorrhoea in lactating women is directly related to both the frequency (Delvoye *et al.*, 1978; McNeilly *et al.*, 1980a; Howie *et al.*, 1981) and duration of suckling (McNeilly *et al.*, 1980a; Howie *et al.*, 1981). Ovarian activity only resumes when suckling activity declines (McNeilly *et al.*, 1980a; Howie *et al.*, 1981, 1982a, b) and is not associated with any specific change in plasma levels of FSH. During the time of ovarian suppression, plasma levels of LH are below those seen in the normal menstrual cycle (Glasier *et al.*, 1983), although pulsatile secretion of LH occurs around 30% of the time even though follicular development is attenuated (Glasier *et al.*, 1984a).

The resumption of ovarian activity is characterized by a high incidence of

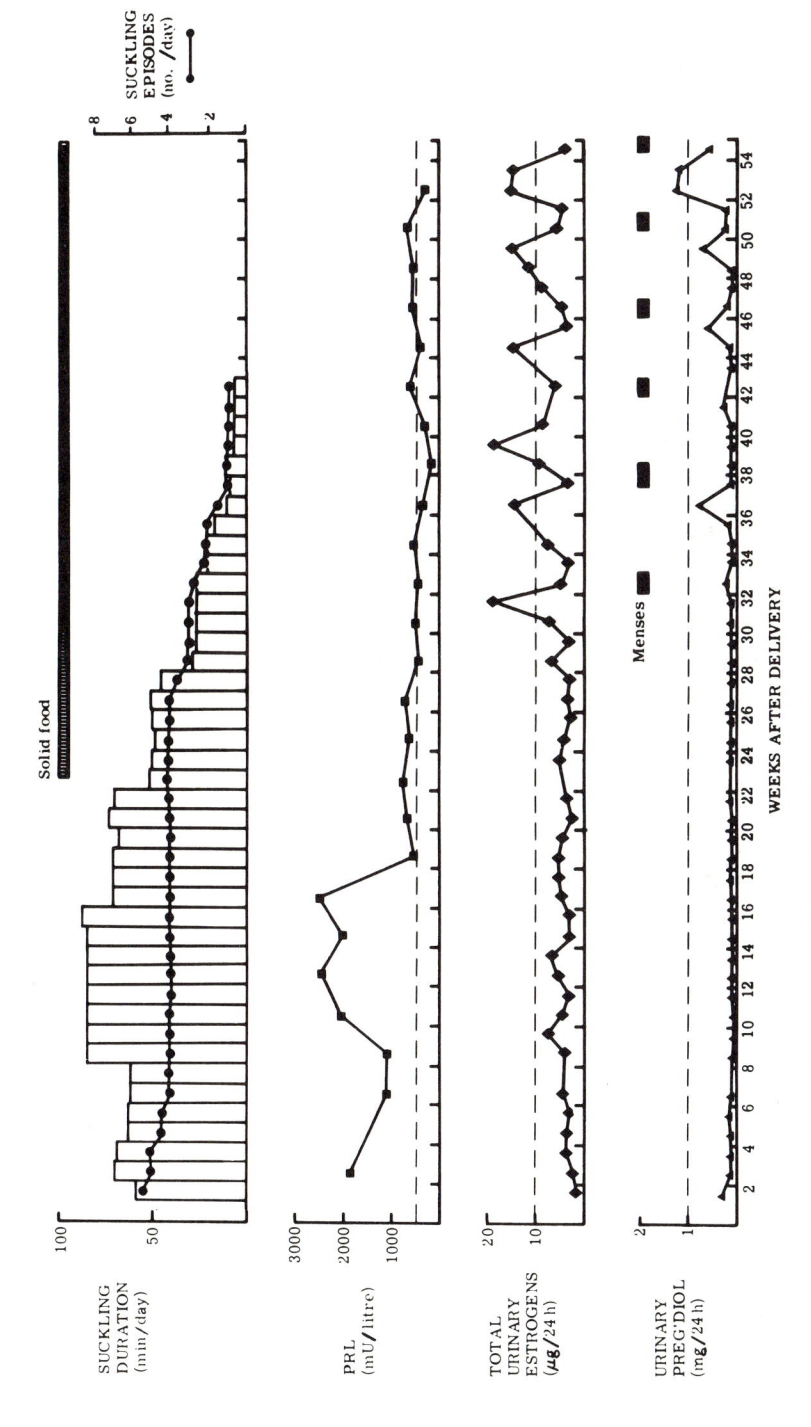

Figure 5. Postpartum changes in urinary total oestrogens and pregnanediol in relation to the sucking frequency, and duration and plasma levels of prolactin in a breast-feeding woman. Note the series of menstrual cycles with inadequate corpus luteum function

luteal phases with inadequate corpus luteum function which improve progressively as suckling activity and plasma levels of prolactin decline (McNeilly et al., 1982b) (Figure 5). A similar progressive improvement in corpus luteum function occurs post partum in women who do not breast feed and in whom plasma levels of prolactin return to normal around the time of the first ovulatory cycle (McNeilly et al., 1980a). Thus it would seem unlikely that elevated levels of prolactin are directly responsible for the formation of these inadequate corpora lutea. More recently it has been shown that there is a gradual increase in the amount of LH released during the preovulatory surge as luteal function improves (Glasier et al., 1984b). Thus it is possible that the inadequacy of corpus luteum function during the resumption of ovulatory cycles in breast feeding women may be more directly related to a deficiency in LH secretion than the elevated levels of prolactin (Glasier et al., 1984b; McNeilly, 1982). On the other hand, the high levels of prolactin may be involved at the hypothalamic pituitary level in the suppression of LH secretion.

It is also surprising that the administration of pulsatile LHRH at a frequency and dose which will induce normal ovulatory cycles in hypogonadotrophic women resulted in the formation of an inadequate corpus luteum in breast-feeding women (A.F. Glasier, A.S. McNeilly and D.T. Baird, unpublished observations). Thus it is possible that a combination of elevated plasma levels of prolactin and lower than normal plasma levels of LH could be responsible for the formation of the inadequate corpora lutea seen in breast-feeding women post partum. The evidence would strongly suggest that hyperprolactinaemia *per se* does not adversely affect corpus luteum function once the corpus luteum has been formed.

5. Conclusions

There is abundant evidence that prolactin is essential for the induction and maintenance of corpus luteum function in rodents. The evidence in other species is scant but would suggest that a similar situation may apply although only minimal levels of prolactin are necessary.

There is little or no evidence that high plasma levels of prolactin *per se* directly inhibit progesterone secretion by the human corpus luteum either in pathological or physiological hyperprolactinaemia.

References

Anderson, L.L. (1966). 'Pituitary–ovarian–uterine relationships in pigs.' *J. Reprod. Fertil. Suppl.*, **1**, 21–32.

Armstrong, D.T., Knudsen, K.A., and Miller, L.S. (1970). 'Effects of prolactin upon cholesterol metabolism and progesterone biosynthesis in corpora lutea of rats hypophysectomized during pseudopregnancy.' *Endocrinology*, **86**, 634–641.

Asch, R.H., Abou-Samra, M., Braunstein, G.D., and Pauerstein, C.J. (1982). 'Luteal

function in hypophysectomized rhesus monkeys.' *J. Clin. Endocrinol. Metab.*, **55**, 154–161.
Bennink, H.J. (1979). 'Intermittent bromocriptine therapy for the induction of ovulation in hyperprolactinaemic patients.' *Fertil. Steril.*, **31**, 267–272.
Butler, W.R., Krey, L.C., Lu, K.-H., Peckham, W.D., and Knobil, E. (1975). 'Surgical disconnection of the medial basal hypothalamus and pituitary function in the Rhesus monkey. IV. Prolactin secretion.' *Endocrinology*, **96**, 1099–1105.
Casper, R.F., and Erickson, G.F. (1981). '*In vitro* heteroregulation of LH receptors by prolactin and FSH in rat granulosa cells.' *Mol. Cell. Endocrinol.*, **23**, 161–171.
Cheng, K.W. (1976). 'Changes in rat ovaries of specific binding for LH, FSH and prolactin during the oestrous cycle and pregnancy.' *J. Reprod. Fertil.*, **48**, 129–135.
Day, S,L., Kirchick, H.J., and Birnbaumer, L. (1980). 'Effect of prolactin on luteal functions in the cyclic rat: positive correlation between luteinizing hormone-stimulated adenylyl cyclase activity and progesterone secretion; role in corpus luteum rescue of the morning surge of prolactin on day 3 of pseudopregnancy.' *Endocrinology*, **106**, 1265–1269.
de Greef, W.F., and Zeilmaker, G.H. (1978). 'Regulation of prolactin secretion during the luteal phase in the rat.' *Endocrinology*, **102**, 1190–1198.
Delpozo, E., Goldstein, M., Friesen, H.G., Brun del Re, R., and Eppenberger, U. (1975). 'Lack of action of prolactin suppression on the regulation of the human menstrual cycle.' *Am. J. Obstet. Gynecol.*, **123**, 719–723.
Delvoye, P., Demaegd, M., Uwayitu, M., and Robyn, C. (1978). 'Serum prolactin, gonadotropins and estradiol in menstruating and amenorrheic women during two years' lactation.' *Am. J. Obstet. Gynecol.*, **130**, 635–640.
Denamur, R., Martinet, J., and Short, R.V. (1966). 'Secretion de la progesterone par les corps jaunes de la brebis après hypophysectomie, section de la tige pituitaire et hysterectomie.' *Acta Endocrinol (Copenhagen)*, **52**, 72–77.
Denamur, R., Martinet, J., and Short, R.V. (1973). 'Pituitary control of the ovine corpus luteum.' *J. Reprod. Fertil.*, **32**, 207–218.
Djahanbakhch, O., McNeilly, A.S., Warner, P.A., Swanston, I.A., and Baird, D.T. (1984). 'Changes in plasma levels of prolactin in relation to those of FSH, oestradiol, androstenedione and progesterone around the preovulatory surge of LH in women.' *Clin. Endocrinol.*, **20**, 463–472.
Dominic, C.J. (1966). 'Effects of single eptopic grafts on the oestrous cycle of the intact mouse.' *J. Reprod. Fertil.*, **12**, 533–538.
Donaldson, L.E., Hansel, W., and Van Vleck, L.D. (1965). 'The luteotropic properties of luteinizing hormone and the nature of oxytocin induced luteal inhibition in cattle.' *J. Dairy Sci.*, **48**, 331–337.
Duchen, M.R., and McNeilly, A.S. (1980). 'Hyperprolactinaemia and long-term lactational amenorrhoea.' *Clin. Endocrinol.*, **12**, 621–627.
Dunaif, A.A., Zimmerman, E.A., Friesen, H.G., and Frantz, A.G. (1982). 'Intracellular localization of prolactin receptor and prolactin in the rat ovary by immunocytochemistry.' *Endocrinology*, **110**, 1465–1471.
Edwards, W.L., Haynes, S.P., and Wilson, J.D. (1982). 'Lack of suppressive effect of prolactin on progesterone production by human granulosa cells *in vitro*.' *Proc. Aust. Soc. Endocrinol.*, **25**, 72.
Ehara, Y., Siler, T., Vandenberg, G., Sinha, Y.N., and Yen, S.S.C. (1973). 'Circulating prolactin levels during the menstrual cycle: episodic release and diurnal variation.' *Am. J. Obstet. Gynecol.*, **117**, 962–970.
Espinosa-Campos, J., Butler, W.R., and Knobil, E. (1975). 'Inhibition of prolactin secretion in the rhesus monkey.' *57th Annual Meeting of the American Endocrine Society*, Abstract 63.

Franchimont, P., Dourcy, C., Legros, J.J., Reuterm, A., Vrindts-Geraert, V., van Cauwenberge, J.R., and Gapard, U. (1976). 'Prolactin levels during the menstrual cycle.' *Clin. Endocrinol.*, **5**, 643–650.

Freeman, M.E., Smith, M.S., Nazian, S.J., and Neill, J.D., (1974). 'Ovarian and hypothalamic control of the daily surges of prolactin secretion during pseudopregnancy in the rat.' *Endocrinology*, **94**, 875–882.

Garris, D.R., and Rothchild, I. (1980). 'Temporal aspects of the involvement of the uterus and prolactin in the establishment of luteinizing hormone-dependent progesterone secretion in the rat.' *Endocrinology*, **107**, 1112–1116.

Garris, D.R., Nanes, M.S., Seguin, C., Kelley, P.A., and Rothchild, I. (1980). 'The lack of relationship between luteinizing hormone (LH) receptors in the rat corpus luteum and the critical need for LH in the luteotropic process.' *Endocrinology*, **107**, 486–490.

Gemzell, C.A. (1975). 'Induction of ovulation in infertile women with pituitary tumors.' *Am. J. Obstet. Gynecol.*, **121**, 311–319.

Gibori, G., and Richards, J.S.C. (1978). 'Dissociation of two distinct luteotropic effects of prolactin regulation of luteinizing hormone-receptor content and progesterone secretion during pregnancy.' *Endocrinology*, **102**, 767–774.

Gibori, G., Richards, J.S., and Keyes, P.L. (1979). 'Synergistic effects of prolactin and estradiol in the luteotropic process in the pregnant rat: regulation of estradiol receptor by prolactin.' *Biol. Reprod.*, **21**, 419–423.

Glasier, A., McNeilly, A.S., and Howie, P.W. (1983). 'Fertility after childbirth: changes in serum gonadotrophin levels in bottle and breast feeding women.' *Clin. Endocrinol.*, **19**, 493–501.

Glasier, A., McNeilly, A.S., and Howie, P.W. (1984a). 'Pulsatile secretion of LH in relation to the resumption of ovarian activity post partum.' *Clin. Endocrinol.*, **20**, 415–426.

Glasier, A.F., McNeilly, A.S., and Howie, P.W. (1984b). 'Changes in urinary steroids and gonadotrophins during early post-partum menstrual cycles in lactating women.' *J. Reprod. Fertil.*, in press.

Gorospe, W.C., and Freeman, M.E. (1981). 'An ovarian role in prolonging and terminating the two surges of prolactin in pseudo-pregnant rats.' *Endocrinology*, **108**, 1293–1298.

Greenwald, G.S. (1967). 'Luteotropic complex of the hamster.' *Endocrinology*, **80**, 118–130.

Hashimoto, I., and Wiest, W.G. (1969). 'Luteotrophic and luteolytic mechanisms in rat corpora lutea.' *Endocrinology*, **84**, 886–892.

Hoffman, B., Schams, D., Bopp, R., Ender, M.L., Giménez, T., and Karg, H. (1974). 'Luteotrophic factors in the cow: evidence for LH rather than prolactin.' *J. Reprod. Fertil.*, **40**, 77–85.

Holt, J.A., Richards, J.S., Midgley, A.R., Jr., and Reichert, L.E., Jr. (1976). 'Effect of prolactin LH receptor in rat luteal cells.' *Endocrinology*, **98**, 1005–1013.

Howie, P.W., McNeilly, A.S., Houston, M.J., Cook, A., and Boyle, H. (1981). 'Effect of supplementary food on suckling patterns and ovarian activity during lactation.' *Br. Med. J.*, **283**, 757–759.

Howie, P.W., McNeilly, A.S., Houston, M.J., Cook, A., and Boyle, H. (1982a). 'Fertility after childbirth: infant feeding patterns, basal PRL levels and post partum ovulation.' *Clin. Endocrinol.*, **17**, 315–322.

Howie, P.W., McNeilly, A.S., Houston, M.J., Cook, A., and Boyle, H. (1982b). 'Fertility after childbirth: postpartum ovulation and menstruation in bottle and breast feeding mothers.' *Clin. Endocrinol.*, **17**, 323–332.

Hunter, M.G. (1980). 'Studies on the corpus luteum *in vitro*.' *Ph.D. Thesis*, University of Edinburgh.

Judd, S.J., Rakoff, J.S., and Yen, S.S.C. (1978). 'Inhibition of gonadotropin and prolactin release by dopamine effect of endogenous estradiol levels.' *J. Clin. Endocrinol. Metab.*, **47**, 494–498.

Kann, G., and Denamur, R. (1974). 'Possible role of prolactin during the oestrous cycle and gestation in the ewe.' *J. Reprod. Fertil.*, **39**, 473–483.

Karsch, F.J., Cook, B., Ellicott, A.R., Foster, D.L., Jackson, G.L., and Nalbandor, A.V. (1971). 'Failure of infused prolactin to prolong the lifespan of the corpus luteum in the ewe.' *Endocrinology*, **89**, 272–275.

Klenscke, H.G., and Brinkley, H.J. (1980a). 'Endogenous rhythms of luteal and adrenal cholesteral ester hydrolase and serum PRL, LH and progesterone in mature pseudopregnant rats.' *Biol. Reprod.*, **22**, 1022–1028.

Klenscke, H.G., and Brinkley, H.J. (1980b). 'Effects of bromocriptine and PRL on luteal and adrenal cholesterol ester hydrolase and serum progesterone concentrations in mature pseudopregnant rats.' *Biol. Reprod.*, **22**, 1029–1039.

Labhetswar, A.P., and Watson, D.J. (1974). 'Temporal relationship between secretory patterns of gonadotropins, estrogens and prostaglandin-F in periparturient rats.' *Biol. Reprod.*, **10**, 103–110.

Lacy, L.R., Knudson, M.M., Williams, J.A., Richards, F.J.S., and Midgley, A.R., Jr. (1976). 'Progesterone metabolism by the ovary of the pregnant rat. Discrepancies in the catabolic regulation model.' *Endicrinology*, **99**, 929–934.

Lamming, G.E. (1978). 'Reproduction during lactation.' In *Control of Ovulation* (eds. D.B. Crighton, N.B. Haynes, G.R. Foxcroft and G.E. Lamming), pp. 335–353. Butterworth, London.

Lenton, E.A., Sulaiman, R., Sobowale, O., and Cooke, I.D. (1982). 'The human menstrual cycle: plasma concentration of prolactin, LH, FSH, oestradiol and progesterone in conceiving and non-conceiving women.' *J. Reprod. Fertil.*, **65**, 131–139.

Louw, B.P., Lishman, A.W., Botha, W.A., and Baumgartner, J.P. (1974). 'Failure to demonstrate a role for the acute release of prolactin at oestrus in the ewe.' *J. Reprod. Fertil.*, **40**, 455–458.

McCracken, J.A., Baird, D.T., and Goding, J.R. (1971). 'Factors affecting the secretion of steroids from the transplanted ovary in the sheep.' *Recent Prog. Horm. Res.*, **27**, 537–647.

McNatty, K.P., Sawers, R.S., and McNeilly, A.S. (1974). 'A possible role for prolactin in control of a steroid secretion by the human Graafian follicle.' *Nature*, **250**, 653–655.

McNeilly, A.S. (1979). 'Effects of lactation on fertility.' *Br. Med. Bull.*, **35**, 151–154.

McNeilly, A.S. (1980). 'Prolactin and the control of gonadotrophin secretion in the female.' *J. Reprod. Fertil.*, **58**, 537–549.

McNeilly, A.S. (1982). 'Prolactin control of ovarian function.' In *Prolactin, Neurotransmission et Fertilité* (eds. H. Clausier and J.-P. Gautray), pp. 1–7. Masson, Paris.

McNeilly, A.S. (1984). 'Prolactin and ovarian function.' In *Neuroendocrine Perspectives* (eds. E.E. Müller and R.M. MacLeod). Elsevier Biomedical Press, Amsterdam (in press).

McNeilly, A.S., and Chard, T. (1974). 'Circulating levels of prolactin during the menstrual cycle.' *Clin. Endocrinol.*, **3**, 105–112.

McNeilly, A.S., Howie, P.W., and Houston, M.J. (1980a). 'Relationship of feeding patterns, prolactin and resumption of ovulation post-partum.' In *Research Frontiers in Fertility Regulation*, (eds. G.I. Zatuchni, M.H. Labbok and J.J. Sciarra), pp. 102–116. Harper & Row, Mexico City.

McNeilly, A.S., Kerin, J., Swanston, I.A., Bramley, T.A., and Baird, D.T. (1980b). 'Changes in the binding of human chorionic gonadotrophin/luteinizing hormone, follicle stimulating hormone and prolactin to human corpora lutea during the menstrual cycle and pregnancy.' *J. Endocrinol.*, **87**, 315–325.

McNeilly, A.S., Abbot, D.H., Lunn, S.F., Chambers, P.C., and Hearn, J.P. (1981). 'Plasma prolactin concentrations during the ovarian cycle and lactation and their relationship to return of fertility post partum in the common marmoset (*Callithrix jacchus*).' *J. Reprod. Fertil.*, **62**, 353–360.

McNeilly, A.S., Glasier, A., Jonassen, J., and Howie, P.W. (1982a). 'Evidence for direct inhibition of ovarian function by prolactin.' *J. Reprod. Fertil.*, **65**, 559–569.

McNeilly, A.S., Howie, P.W., Houston, M.J., Cook, A., and Boyle, H. (1982b). 'Fertility after childbirth: adequacy of post-partum luteal phases.' *Clin. Endocrinol.*, **17**, 609–615.

Milligan, S.R., and MacKinnon, P.C.B. (1976). 'Correlation of plasma LH and prolactin levels with the fate of the corpus luteum in the vole, *Microtus agrestis*.' *J. Reprod. Fertil.*, **47**, 111–113.

Morishige, W.T., and Rothchild, I. (1974). 'Temporal aspects of the regulation of corpus luteum function by luteinizing hormone, prolactin and placental luteotrophin during the first half of pregnancy in the rat.' *Endocrinology*, **95**, 260–274.

Murakami, N., Takahashi, M., Suzuki, Y., and Homma, K. (1982). 'Responsiveness of dispered rat luteal cells to luteinizing hormone and prolactin during the estrous cycle and early pseudopregnancy.' *Endocrinology*, **111**, 500–508.

Murphy, B.D. (1979). 'The role of prolactin in implantation and luteal maintenance in the ferret.' *Biol. Reprod.*, **21**, 517–521.

Nadler, R.D., Graham, C.E., Collins, D.C., and Gould, K.G. (1979). 'Plasma gonadotropins, prolactin, gonadal steroids and genital swelling during the menstrual cycle of Lowland Gorillas.' *Endocrinology*, **105**, 290–296.

Niswender, G.D. (1974). 'Influence of 2-Br-α-ergocryptine on serum levels of prolactin and the oestrous cycle in sheep.' *Endocrinology*, **94**, 612–615.

Nolin, J.M. (1980). 'Incorporation of endogenous prolactin by granulosa cells and dictyate oocytes in the post partum rat: effects of estrogen.' *Biol. Reprod.*, **22**, 417–422.

Oxberry, B.A., and Greenwald, G.S. (1982). 'An autoradiographic study of the binding of ^{125}I-labelled follicle-stimulating hormone, human chorionic gonadotropin and prolactin to the hamster ovary throughout the estrous cycle.' *Biol. Reprod.*, **27**, 505–516.

Reyes, F.I., Winter, J.S.D., Faiman, C., and Hobson, W.C. (1975). 'Serial serum levels of gonadotrophins, prolactin and sex steroids in non-pregnant and pregnant chimpanzee.' *Endocrinology*, **96**, 1447.

Reyes, F.I., Gomez, F., and Faiman, C. (1977). 'Pathological hyperprolactinaemia: a 5 year experience.' In *Prolactin and Human Reproduction* (eds. P.G. Crosignani and C. Robyn) pp. 71–96. Academic Press, New York.

Richards, J.S., and Williams, J.J. (1976). 'Luteal cell receptor content for prolactin (PRL) and luteinizing hormone (LH): regulation by LH and PRL.' *Endocrinology*, **99**, 1571–1581.

Richardson, M.C., and Masson, G.M. (1981). 'Stimulation of human chorionic gondotrophin of oestradiol production by dispersed cells from human corpus luteum: comparison with progesterone production; utilization of exogenous testosterone.'. *J. Endocrinol.*, **91**, 197–203.

Robyn, C., Delvoy, P., Van Exter, C., Vekumans, M., Caufriez, A., De Nayer, P., Delogne-Desnoek, J., and l'Hermite, M. (1977). 'Physiological and pharmacological factors influencing prolactin secretion and their relation to human reproduction.' In *Prolactin and Human Reproduction* (eds. P.G. Crosighani and C. Robyn), pp. 71–96. Academic Press, New York.

Rolland, R., Gunsalus, G.L., and Hammond, J.M. (1976). 'Demonstration of specific binding of prolactin by porcine corpora lutea.' *Endocrinology*, **98**, 1083–1091.

Rothchild, I. (1981). 'The regulation of the mammalian corpus luteum.' *Recent Prog. Horm. Res.*, **37**, 183–298.
Saito, T., and Saxena, B.B. (1975). 'Specific receptors for prolactin in the ovary.' *Acta Endocrinol.*, **80**, 126–137.
Sarris, S., Dwyer, G.I.M., McGarrigle, H.H.G., Lawrence, D.M., Little, V., and Lachelin, G.C.L. (1978). 'Prolactin and luteal insufficiency.' *Clin. Endocrinol.*, **9**, 243–247.
Schulz, K.D., Geiger, W., Del Pozo, E., and Kunzig, H.J. (1978). 'Pattern of sexual steroids, prolactin and gonadotropic hormones during prolactin inhibition in normally cycling women.' *Am. J. Obstet. Gynecol.*, **132**, 561–566.
Seppala, M., Hirvonen, E., and Ranta, T. (1976). 'Hyperprolactinaemia and luteal insufficiency.' *Lancet*, **i**, 229–230.
Sernia, C., and Tyndale-Biscoe, C.H. (1979). 'Prolactin receptors in the mammary gland, corpus luteum and other tissues of the Tammar Wallaby, *Macropus eugenii*.' *J. Endocrinol.*, **83**, 79–89.
Smith, M.S., and Neill, J.D. (1976). 'Termination at mid pregnancy of the two daily surges of plasma prolactin initiated by mating in the rat.' *Endocrinology*, **98**, 696–701.
Smith, M.S., McClean, B.K., and Neill, J.D. (1976). 'Prolactin: the initial luteotropic stimulus of pseudopregnancy in the rat.' *Endocrinology*, **98**, 1370–1377.
Smith, W.R., McShan, T.H., and Casida, L.I.E. (1957). 'On maintenance of the corpora lutea of the bovine with lactogen.' *J. Dairy Sci.*, **40**, 443.
Stouffer, R.L., Coensgen, J.L., and Hodgen, G.D. (1980). 'Progesterone production by luteal cells during acute incubation and cell culture.' *Steroids*, **34**, 523–532.
Swanston, I.A., Djahanbahkch, O., and McNeilly, A.S. (1984). 'The effect of prolactin on human pre-ovulatory granulosa cells maintained in culture.' *Acta Endocrinol.*, (in press).
Tann, G.J.S., and Biggs, J.S.G. (1983). 'Effects of prolactin on steroid production by human luteal cells *in vitro*.' *J. Endocrinol.*, **96**, 499–503.
Thorner, M.O. (1977). 'Prolactin'. *Clin. Endocrinol. Metab.*, **6**, 201–222.
Vanrell, J.A., and Balasch, J. (1983). 'Prolactin in the evaluation of luteal phase infertility.' *Fertil. Steril.*, **39**, 30–34.
Weiss, G., Butler, W.R., Hotchkiss, J., Dierschke, D.J., and Knobil, E. (1976). 'Periparturitional serum concentrations of prolactin, the gonadotrophins and the gonadal hormones in the Rhesus monkey.' *Proc. Soc. Exp. Biol. Med.*, **151**, 113–116.
Wiest, W.G., Kidwell, W.R., and Baloch, K., Jr. (1968). 'Progesterone catabolism in the rat ovary: a regulatory mechanism for progesterone potency during pregnancy.' *Endocrinology*, **82**, 844–859.
Wu, D.H., Wiest, W.G., and Enders, A.C. (1976). 'Luteotropic regulation of dispersed rat luteal cells in early pregnancy.' *Endocrinology*, **98**, 1378–1389.

The Luteal Phase
Edited by S.L. Jeffcoate
© 1985 John Wiley & Sons Ltd.

CHAPTER 6

Luteal function after ovulation induction by pulsatile luteinizing hormone releasing hormone

S. Franks, Z. van der Spuy, W.P. Mason, J. Adams and H.S. Jacobs

*Department of Obstetrics and Gynaecology,
St. Mary's Hospital Medical School,
London W2 1PG, and
The Middlesex Hospital,
London, W1 8AA, UK*

Introduction

Infertility associated with anovulation is usually due to disordered regulation of gonadotrophins, and drugs that are able to restore the normal pattern of endogenous gonadotrophin secretion (e.g. clomiphene or bromocriptine) or to stimulate the ovary directly (exogenous gonadotrophins) have been used successfully to induce ovulation. An important recent development in treatment of hypogonadotrophic women is the use of long-term, low-dose therapy with luteinizing hormone releasing hormone (LHRH) administered by pulsatile infusion pump (Crowley and McArthur, 1980; Leyendecker et al., 1980; Mason et al., 1984). This treatment is able to reproduce the pattern of gonadotrophin and ovarian steroid secretion observed during a normal ovulatory menstrual cycle with little danger of hyperstimulation of the ovary, which is the principal disadvantage of exogenous therapy with human menopausal gonadotrophin (hMG) and human chorionic gonadotrophin (hCG).

After induction of ovulation with clomiphene, the range of serum progesterone concentrations during the luteal phase tends to be rather wider than is found in spontaneous cycles (Hull et al., 1982); this probably reflects progesterone production from more than one (but rarely more than two)

postovulatory follicles. The common occurrence of multiple follicular development and multiple ovulation following treatment with hMG and hCG accounts for the very high serum progesterone concentrations that are frequently observed during the luteal phase of gonadotrophin-induced cycles. Treatment with hMG and hCG may induce premature ovulation of immature Graafian follicles which then become luteinized. For this reason measurement of serum progesterone concentrations during the luteal phase of gonadotrophin-induced cycles does not predict, reliably, the chance of conception (Hull *et al.*, 1982).

The treatment of choice for induction of ovulation in women with hyperprolactinaemic anovulation is a dopamine agonist such as bromocriptine. This drug lowers prolactin levels and induces normal gonadal function (Franks *et al.*, 1977). Luteal function in these women is indistinguishable from that in spontaneous cycles, apart from during the first few weeks of treatment when short or inadequate luteal phases have been reported in some patients (Moult *et al.*, 1982).

In this chapter we shall describe the function of the corpus luteum in ovulatory cycles induced in hypogonadotrophic women by LHRH. We shall show how manipulation of LHRH therapy during the luteal phase may provide important information about gonadotrophic dependency of the human corpus luteum, and we shall discuss the dynamics of gonadotrophin and progesterone secretion assessed by sequential blood sampling after LHRH-induced ovulation.

1. LHRH Therapy: Methods

The use of pulsatile low-dose LHRH treatment to induce ovulatory cycles in women with hypogonadotrophic hypogonadism was described by Crowley and McArthur (1980) and was the logical sequel to the classical series of experiments in the LHRH-deficient Rhesus monkey performed by Knobil (1980). Our own group has recently reported successful induction of ovulation in 128 cycles in 28 hypogonadotrophic patients (Mason *et al.*, 1984) using a small automatic pulsatile infusion pump designed for us by Dr. Ian Sutherland and his colleagues at the National Institute of Medical Research, Mill Hill. This pump is a modification of the miniature insulin infuser described by Rothwell *et al.* (1983). LHRH (HRF from Ayerst or HOE-471 from Hoechst) was administered at a dose of 15 μg every 90 min throughout the cycle. In most cases (30 of 36 cycles) the hormone was given by the subcutaneous route, and in the remainder intravenously. Data were analysed from 16 conception cycles and 20 ovulatory non-conception cycles in a total of 22 women. Follicular growth, ovulation and luteal function were monitored by pelvic ultrasound and measurement of gonadotrophins, oestradiol and progesterone in serum on alternate days throughout the cycle.

2. Luteal Function in Non-conception and Conception Cycles

2.1. *Progesterone Secretion and Relation to Follicular Growth*

Serum progesterone concentrations in 20 ovulatory, non-conception cycles and 17 conception cycles are shown in Figure 1. The mean length of the luteal phase in non-conception cycles (from the peak of the mid-cycle LH surge to onset of menses) was 13.4 ± 0.8 (SD) days; the pattern of progesterone secretion in these patients was indistinguishable from that in normal spontaneous menstrual cycles (see Chapter 7). In contrast to the luteal phase after treatment with clomiphene of hMG/hCG, serum progesterone levels were within the physiological range throughout. It is of considerable interest that the luteal phase in LHRH-induced cycles resembles normal corpus luteum function so closely despite what may be regarded as a non-physiological LHRH stimulus. The pulse frequency of LH release in spontaneous cycles diminishes during the luteal phase to a rate of approximately one pulse per 180 min

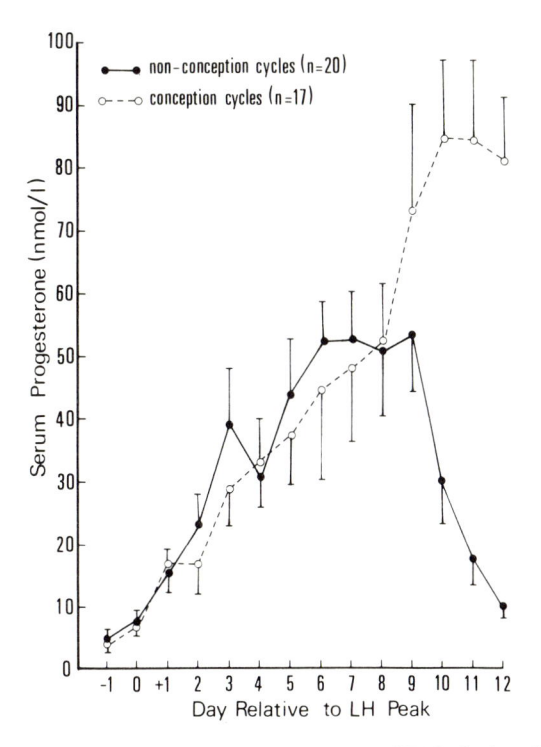

Figure 1. Serum progesterone concentrations (mean + SEM) during the luteal phase of LHRH-induced ovulatory cycles. There was no significant difference between the mean serum progesterone levels in conception ($n = 17$) and non-conception ($n = 20$) cycles until day +9

(Backstrom *et al.,* 1982). In the LHRH-induced cycles in our patients the pulse rate of one per 90 min was maintained throughout the luteal phase but resulted in a normal pattern of progesterone secretion (and, indeed, serum LH concentrations; see below), suggesting that progesterone may modulate the response of LH to LHRH at the pituitary level. Further evidence for this phenomenon is provided by analysis of sequential hormone measurements during LHRH therapy (section 4).

There has been some controversy about whether progesterone secretion in the early luteal phase of conception cycles is different from that in ovulatory non-conception cycles (Chapter 7). Cooke and co-workers have suggested that concentrations of progesterone in this phase of the cycle are higher in cycles in which conception occurs and that this reflects differences in follicular maturation between conception and non-conception cycles. In our homogeneous group of subjects in whom ovulation was induced by LHRH, we found no significant difference between progesterone levels in conception and non-conception cycles at any point up to day +9 from the mid-cycle surge (Figure 1). We also examined follicular growth in the two groups; the results are shown in Figure 2. Growth of the preovulatory follicle, as determined by ultrasound assessment of follicular diameter, was identical in conception and non-conception cycles. Figure 3 shows the correlation between follicular diameter and serum oestradiol concentrations in LHRH-induced cycles, confirming the findings in spontaneous cycles (Hackeloer *et al.,* 1979). Thus there is no difference between non-conception and conception cycles in the function of the

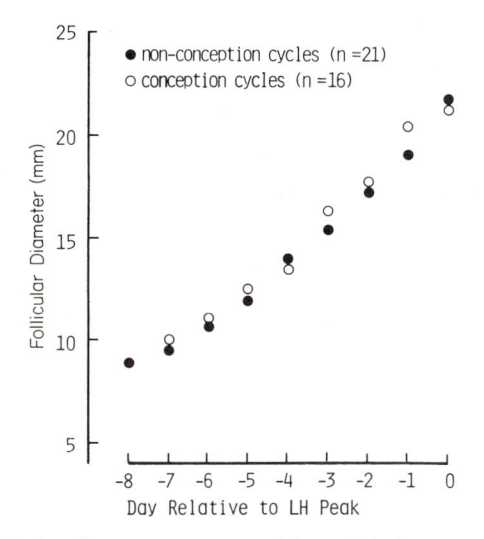

Figure 2. Mean follicular diameter, as assessed by serial ultrasound measurements, in ovulatory cycles induced by LHRH therapy. The error bars are omitted for the sake of clarity. There was no difference between conception and non-conception cycles

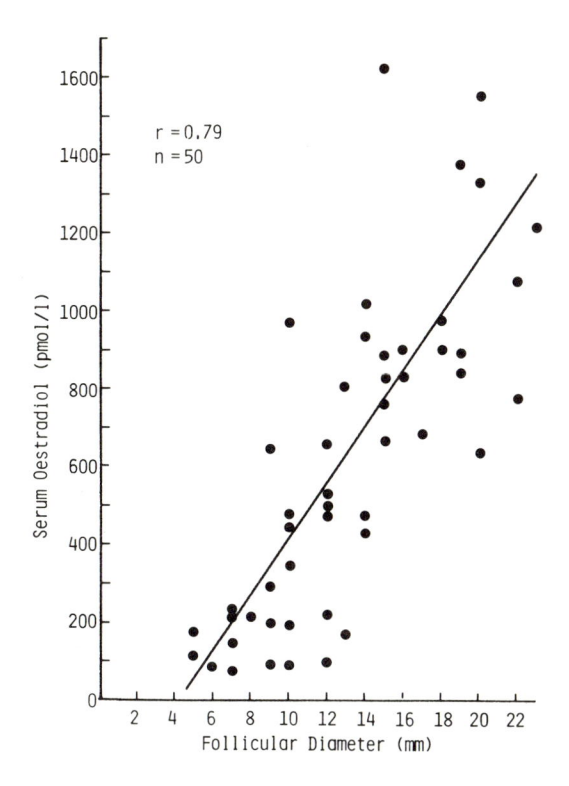

Figure 3. Correlation of follicular diameter and serum oestradiol measurements in LHRH-induced ovulatory cycles. $y = 74x - 311$, $r = 0.79$. $p < 0.0001$

Graafian follicle as judged from ultrasound assessment of preovulatory growth and postovulatory progesterone secretion.

2.2 Serum Oestradiol and Gonadotrophin Concentrations in the Luteal Phase

Serum concentration of oestradiol, LH and follicle-stimulating hormone (FSH) in the luteal phase of LHRH-induced cycles and their relationship to serum progesterone levels are shown in Figure 4. The pattern of secretion of oestradiol by the corpus luteum is similar to that of progesterone, and oestradiol levels rise in parallel to those of progesterone if conception occurs. LH and FSH concentrations mimic those observed in the luteal phase of the normal menstrual cycle (Midgley and Jaffe, 1968) despite what may be considered to be a non-physiological stimulus to the pituitary – the continuation of LHRH therapy at 90-min intervals. In the non-conception cycles it is interesting to note (as in spontaneous cycles) the rise of FSH as progesterone secretion wanes, which

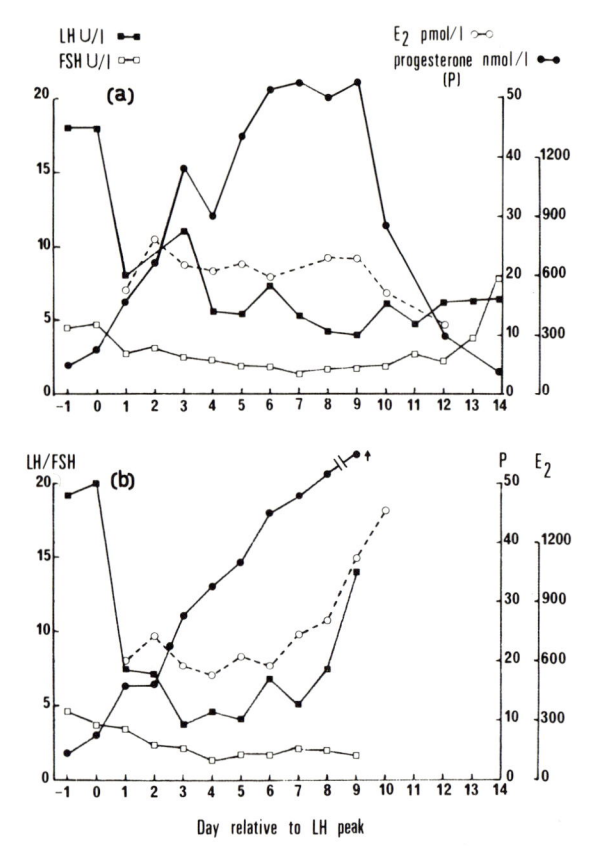

Figure 4. Correlation of gonadotrophins and gonadal steroids in (a) 20 LHRH-induced non-conception cycles and (b) 16 LHRH-induced conception ovulatory cycles. Only the mean values are shown

heralds the onset of menstruation and which is important in initiation of the subsequent cycle.

2.3 Uterine Size in the Luteal Phase of Conception and Non-conception Cycles

In monitoring the response to LHRH therapy we have included ultrasound assessment of uterine size (Adams et al., 1984). Uterine cross-sectional area increases during the follicular phase of the cycle and this correlates well with increasing follicle diameter and with oestrogen production. During the luteal phase, if conception does not occur, there is a gradual decrease in uterine area (Figure 5) but the remarkable finding is the change in uterine size if conception occurs. In conception cycles, uterine size is maintained and indeed increases

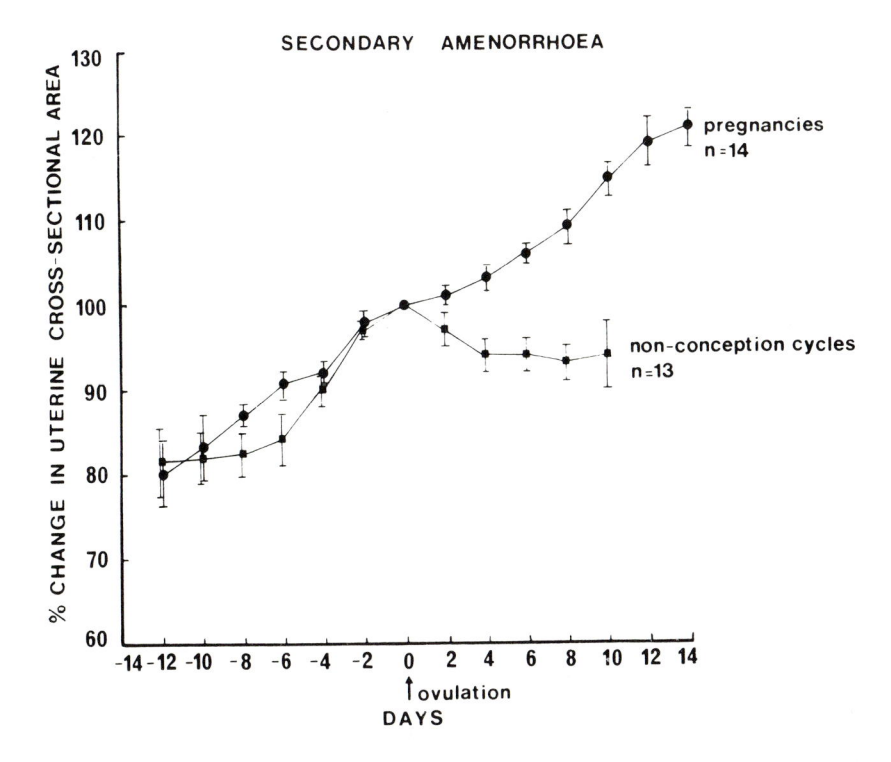

Figure 5. Changes in uterine cross-sectional area (expressed as percentage of area on day of ovulation) in pregnancy and non-conception cycles induced by pulsatile LHRH

during the luteal phase. The difference in uterine area between conception and non-conception cycles is statistically significant as early as 4 days after ovulation (Figure 5). As yet we do not know which hormonal factor is responsible for these changes, but their occurrence so early in pregnancy raises the question of whether the preimplantation blastocyst is hormonally active.

3. Gonadotrophin Dependency of the Corpus Luteum

The degree to which the human corpus luteum is dependent on gonadotrophin support is poorly understood. A single ovulatory injection of hCG in women in whom follicular growth has been induced with hMG may be enough to ensure normal luteal function and conception (Gemzell, 1965; Vande Wiele *et al.*, 1970), but the half-life of hCG in the circulation is much longer than that of LH and when ovulation was triggered by administration of human LH to hypogonadotrophic women progesterone secretion could be maintained for a maximum of 5 days (Vande Wiele *et al.*, 1970). In a recent study, Asch *et al.* (1982) gave hCG or hLH to hypophysectomized Rhesus monkeys and concluded that luteal function could be maintained for up to 10 days after a single preovulatory injection of hLH. However, Hutchinson and Zeleznick (1983), in experiments on LHRH-deficient Rhesus monkeys, showed that, although ovulation could be induced by administration of pulsatile LHRH, LH fell to undetectable levels, progesterone secretion decreased and menses ensued if LHRH therapy was dicontinued in the mid-luteal phase. Our own studies with pulsatile LHRH therapy have enabled us to observe the effects of withdrawal of gonadotrophin support at various stages in the luteal phase. The pump was stopped in three women at 1, 3 and 5 days, respectively, after ovulation (confirmed by ultrasound). A summary of the results is shown in Table 1. In each case growth of the preovulatory follicle, as judged by oestradiol measurement and ultrasound monitoring, had been normal and serum progesterone levels had risen. In each patient stopping the pump resulted in a prompt fall in progesterone and, 2 days later, menstruation. Details of the patient whose pump was removed 1 day after ovulation are shown in Figure 6. Thus, these data suggest that the normal function of the corpus luteum is dependent on gonadotrophin support for at least 5 days after ovulation.

Table 1. Gonadotrophin dependency of the corpus luteum: summary of clinical and endocrine data in three hypogonadotrophic women in whom the pulsatile LHRH was stopped at various times after ovulation (confirmed by ultrasound). Note the occurrence of menstrual bleeding in each case 2 days after withdrawal of LHRH

	Patient 1	Patient 2	Patient 3
Day LHRH stopped (relative to ovulation)	+1	+3	+5
Maximum follicular diameter (mm)	–	20	20
or serum oestradiol (pmol/l)	1627	–	910
Maximum serum progesterone (nmol/l)	25	–	16
Day of menstruation	+3	+5	+7

4. Episodic Secretion of Gonadotrophins in the Luteal Phase

In order to study the pituitary responsiveness to LHRH therapy during the luteal phase, we measured concentrations of LH, FSH and progesterone in sequential blood samples taken at 15-min intervals for 8 h on day 21 of a 28-day cycle in a patient receiving pulsatile LHRH (Figure 7). Progesterone concentrations were within the expected mid-luteal range. We were surprised to observe that there was a significant increase in LH concentration after only three of the five 'pulses' of LHRH and in one of these the change was less than 1 IU/l. Our

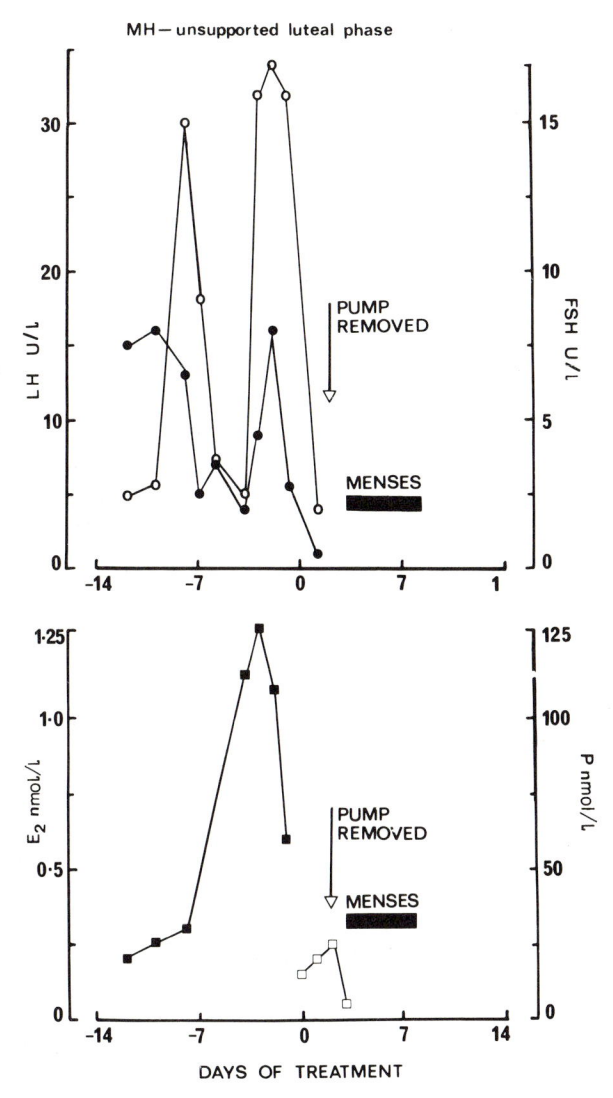

Figure 6. Gonadotrophin and gonadal steroid concentrations in the serum of a patient in whom the LHRH pump was removed 1 day after ovulation (proved by ultrasound). Note the rapid fall in progesterone concentrations (■) and onset of menstruation 2 days later

own studies in normal men have shown that subcutaneous administration of LHRH consistently and reproducibly produces a pulse of immunoreactive LHRH in the peripheral circulation and that this is followed by an increase in LH (Abdul Wahid *et al.*, 1984). Thus it seems unlikely that this inconsistency of the pituitary response to LHRH can be explained by a variation in LHRH

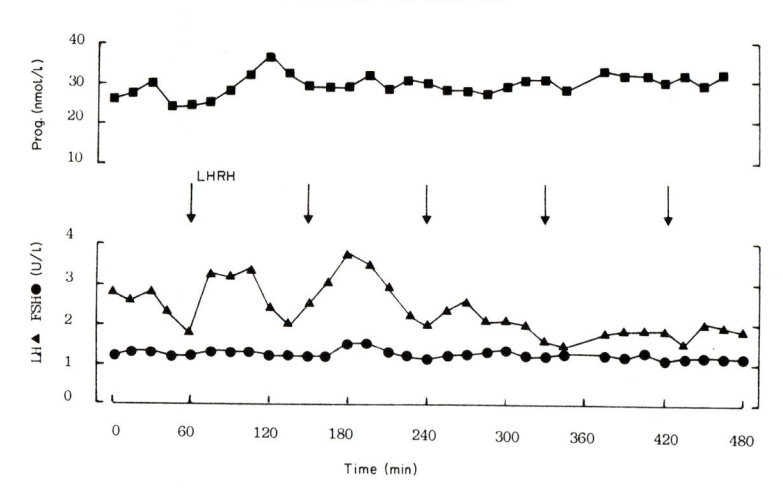

Figure 7. Sequential blood samples (every 15 min for 8 h) in the mid-luteal phase of a patient whose ovulation was induced with LHRH and then continued throughout the cycle. Note the failure of LH and FSH concentrations to rise in response to the final two pulses of LHRH. Serum progesterone (prog) concentrations are in the mid-luteal range

stimulus, especially as we have observed a similar pattern during the luteal phase in a patient receiving intravenous LHRH. It therefore appears that LH secretion in response to a constant signal of LHRH can be modified at pituitary level, presumably by the feedback action of progesterone. This modification may affect the apparent number of LH 'pulses' observed during the sampling period, which suggests that the reduction in LH pulse frequency during the luteal phase observed by Backstrom *et al.* (1982) may involve a pituitary as well as a hypothalamic site action of progesterone.

5. Conclusion

Pulsatile LHRH therapy, which reproduces the physiological changes in the pituitary and ovary, is a successful form of ovulation induction in patients with hypogonadotrophic hypogonadism. We have been able to show that progesterone secretion in LHRH-induced ovulatory cycles is similar to that in spontaneous cycles and that there is no difference in follicular growth and early luteal progesterone concentrations between ovulatory, non-conception cycles and conception cycles. There is, however, a difference in the pattern of uterine growth following conception which is detectable by ultrasound measurement in the early luteal phase and before there is any measurable difference in oestradiol or progesterone concentrations. The hormonal factor (and its source) that is responsible for this very early pregnancy change remains to be elucidated. From our preliminary studies of the effect of withdrawing LHRH therapy at various stages during the luteal phase, we conclude that the human corpus luteum requires the support of gonadotrophins until at least 5 days after ovulation. Finally, our analysis of the episodic nature of LH secretion during pulsatile LHRH therapy suggests that progesterone may have an important modulating effect on the LH response to LHRH at the level of the pituitary during the luteal phase.

References

Abdul-Wahid, N., Morris, D., Jeffcoate, S.L., and Jacobs, H.S. (1984). 'LHRH and LH concentrations in serum after subcutaneous or intravenous administration of LHRH to normal men.' In preparation.

Adams, J., Mason, W.P., Tucker, M., and Jacobs, H.S. (1984). 'Uterine growth in LHRH-induced ovulation cycles.' In preparation.

Asch, R.H., Moustapha, A.S., Braunsein, G.D., and Pauerstein, C.J. (1982). 'Luteal function in hypophysectomized Rhesus monkey.' *J. Clin. Endocrinol. Metab.*, **55**, 154–161.

Backstrom, C.T., McNeilly, A.S., Leask, R.M., and Baird, D.T. (1982). 'Pulsatile secretion of LH, FSH, prolactin, oestradiol and progesterone during the human menstrual cycles.' *Clin. Endocrinol.*, **17**, 29–42.

Crowley, W.F., and McArthur, J.W. (1980). 'Stimulation of the normal menstrual cycle in Kallman's Syndrome by pulsatile administration of luteinizing hormone-releasing hormone (LHRH).' *J. Clin. Endocrinol. Metab.*, **51**, 173–175.

Franks, S., Jacobs, H.S., Hull, M.G.R., Steele, S.J., and Nabarro, J.D.N. (1977). 'Management of hyperprolactinaemic amenorrhoea.' *Br. J. Obstet. Gynaecol.*, **84**, 241–243.

Gemzell, C. (1965). 'Induction of ovulation with human gonadotrophins.' *Recent Prog. Horm. Res.*, **21**, 179–199.

Hackeloer, B.J., Fleming, R., Robinson, H.P., Adam, A.H., and Coutts, J.R.T. (1979). 'Correlation of ultrasonic and endocrinologic assessment of human follicular development.' *Am. J. Obstet. Gynecol.*, **135**, 122–127.

Hull, M.G.R., Savage, P.E., Bromham, D.R., Ismail, A.A.A., and Morris, A.F. (1982). 'The value of a single serum progesterone measurement in the midluteal phase as a criterion of a potentially fertile cycle ("ovulation") derived from treated and untreated conception cycles.' *Fertil. Steril.*, **37**, 355–360.

Hutchinson, J.S., and Zeleznick, A.J. (1983). 'The Rhesus monkey corpus luteum is dependent upon pituitary gonadotrophin secretion throughout the luteal phase of the menstrual cycle.' *Biol. Reprod.*, **28**, (Suppl. 1), Abstract 94.

Knobil, E. (1980). 'The neuroendocrine control of the menstrual cycle.' *Recent Prog. Horm. Res.*, **36**, 53–86.

Leyendecker, G., Wildt, L., and Hansman, M. (1980). 'Pregnancies following chronic intermittent (pulsatile) administration of GnRH by means of a portable pump ("Zyklomat") – a new approach to the treatment of infertility in hypothalamic amenorrhoea.' *J. Clin. Endocrinol. Metab.*, **51**, 1214–1216.

Mason, W.P., Adams, J., Morris, D.V., Tucker, M., Price, J., Wheeler, M.J., Sutherland, I., Chambers, G., White, S., and Jacobs, H.S. (1984). 'Induction of ovulation by pulsatile luteinising hormone-releasing hormone.' *Br. Med. J.*, **288**, 181–185.

Midgley, A.R., and Jaffe, R.B. (1968). 'Regulation of human gonadotrophins. iv. Correlation of serum concentrations of follicle-stimulating and luteinizing hormones during the menstrual cycle.' *J. Clin. Endocrinol. Metab.*, **28**, 1699–1708.

Moult, P.J.A., Rees, L.H., and Besser, G.M. (1982). 'Pulsatile gonadotrophin secretion in hyperprolactinaemic amenorrhoea and response to bromocriptine therapy.' *Clin. Endocrinol.*, **16**, 153–162.

Rothwell, D., Sutherland, I.A., Pickup, J.C., Bending, J.J., Keen, H., and Parsons, J.A. (1983). 'A new miniature, open-loop extra corporeal insulin infusion pump.' *J. Biomed. Eng.*, **5**, 178–184.

Vande Wiele, R.L., Bogumil, J., Dyrenfurth, I., Ferin, M., Jewelewicz, R., Warren, M., Rizkallah, T., and Mikhail, G. (1970). 'Mechanisms regulating the menstrual cycle in women.' *Recent Prog. Horm. Res.*, **26**, 63–92.

The Luteal Phase
Edited by S.L. Jeffcoate
© 1985 John Wiley & Sons Ltd.

CHAPTER 7

The abnormal luteal phase

J. R. T. COUTTS

*Department of Obstetrics and Gynaecology,
The University of Glasgow,
Glasgow,
UK*

Introduction

For many years the abnormal or deficient luteal phase has been allegedly implicated in infertility, particularly in women with apparently ovulatory cycles (unexplained infertility). Many previous diagnoses of abnormal luteal phases were based on basal body temperature charts, luteal phase hormone analyses (usually on isolated blood or urine samples) and apparent luteal phase lengths and were to a large extent poorly documented.

The advent of ovarian ultrasonography (Hackeloer *et al.*, 1978) and the application of sensitive hormone assays to serial daily samples throughout human menstrual cycles have allowed the normal luteal phase (and hence the abnormal luteal phase) to be better defined.

This chapter compares the parameters of normal luteal phase with those in women with unexplained infertility both endocrinologically and ultrasonically and delineates several abnormalities. An attempt is also made to establish whether the luteal phase abnormality with the greatest incidence is a causative factor in such parients' infertility.

1. Volunteers, Patients and Methods

1.1. Normal Volunteers

Serial daily plasma samples were obtained throughout a single menstrual cycle from 12 apparently normally menstruating volunteers. The criteria for 'normality' accepted for these volunteers (aged 18–28 years) who were all nonparous at the time of sampling, were: (i) a history of regular menstrual cycles of from 26 to 32 days' duration; (ii) no history of medical events which had affected their regular cycles during the past 3 years; (iii) no history of oral contraception; (iv) a classical mid-cycle LH peak; and (v) a luteal phase (day of LH peak to day of onset of succeeding menses) of at least 13 days' duration. Nine of these 12 volunteers are now of proven fertility, whilst of the others, none to our knowledge has been attempting to conceive without success.

1.2. Patients

Serial daily plasma samples were also obtained throughout at least one menstrual cycle from 247 female partners of couples with unexplained infertility. Such couples fulfilled the following criteria: (1) the female had regular and normal length menstrual cycles; (2) she had patent tubes and no obvious anatomical anomalies on laparoscopic examination; (3) the husband had normal seminal analyses; and (4) the infertility was of a minimun of 3 years' duration.

1.3. Methods

1.3.1. Hormone analyses All plasma samples were assayed using radio-immunoassays of appropriate sensitivity and specificity (Coutts *et al.*, 1981) to determine the concentrations of luteinizing hormone (LH), follicle-stimulating hormone (FSH), prolactin (PRL), oestradiol (E_2) and progesterone (P).

The day of the LH peak was designated day 0 to allow comparison of cycles of different lengths. Days prior to the LH peak (follicular phase) were given negative signs and days after the LH peak (luteal phase) positive signs. Normal ranges were constructed from the volunteer cycles and used for comparison with the hormone profiles of the cycles from the infertile women.

1.3.2. P index for the early luteal phase For each cycle an index of P concentrations ([P]) was calculated for the early luteal phase using the formula:

$$\text{P Index} = \frac{\Sigma[P] \text{ days } +2 \text{ to } +6 \text{ in the patient}}{\Sigma[P] \text{ days } +2 \text{ to } +6 \text{ of mean normal cycle data}} \times 100$$

A luteal phase profile equivalent to the mean of the normal cycle data therefore had a P index of 100 and the limits of the indices from the normal cycle ranges were 70 and 130. Values below 70 thus indicated subnormal early luteal phase P concentrations.

1.3.3. Ovarian ultrasonography Ovarian ultrasound was performed using the full-bladder technique (Adam, 1981) and an Emisonic 4201 fixed B scanner with a 3.5 MHz medium-focus transducer. Follicular diameter (FD) was defined as the mean of the largest diameter and its bisecting perpendicular diameter observed in both transverse and longitudinal planes. Ovarian ultrasound was performed on at least four occasions during the pre/periovulatory periods and on at least three occasions after day +1.

2. Results and Discussion

2.1. Normal Luteal Phase

The major problem in the evaluation of the *abnormal* luteal phase is the precise definition of the *normal* luteal phase, and the minimum requirements of a luteal phase for maintenance of a conception are unknown. Most groups of workers have constructed normal luteal phase hormone ranges by analyses of serial plasma or serum samples throughout menstrual cycles from apparently normal volunteers. It is now well established that the most reliable profiles derive from daily sampling regimes from volunteers or patients. The ranges (mean ± SD) for E_2 and P obtained in our laboratory from 12 'normal' volunteers who fulfilled the criteria described in section 1 are shown in Figure 1.

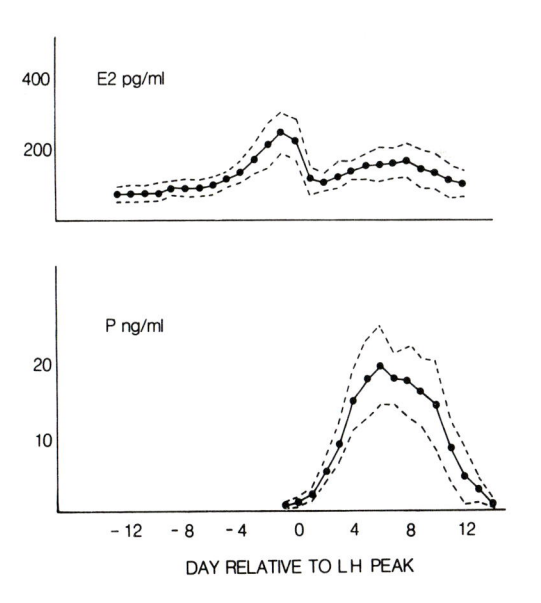

Figure 1. Mean (●) ± SD (– – –) concentrations of P and E_2 from normally cycling volunteers (n = 12) relative to the day of the LH peak. To convert to SI units multiply by 3.3 (P, nmol/l) or 3.7 (E_2, pmol/l)

The usefulness of such ranges obtained from 'normal volunteers' has been questioned by Lenton et al. (1982), who reported that early luteal phase levels of P were significantly higher in conception than in non-conception cycles. We have been unable to confirm these findings. Figure 2 compares our normal cycle ranges for E_2 and P with mean (± SD) profiles from eight spontaneous conception cycles and no differences were observed until pregnancy had been established (day +10). Laufer et al. (1982) also reported no difference between conception and non-conception cycles although only a limited number of luteal phase samples were examined in their study. Perhaps the differences observed by Lenton et al. (1982) were either a function of their selection of control cycles and/or the fact that their conception cycles were not all spontaneous (i.e. may have involved pharmacological agents).

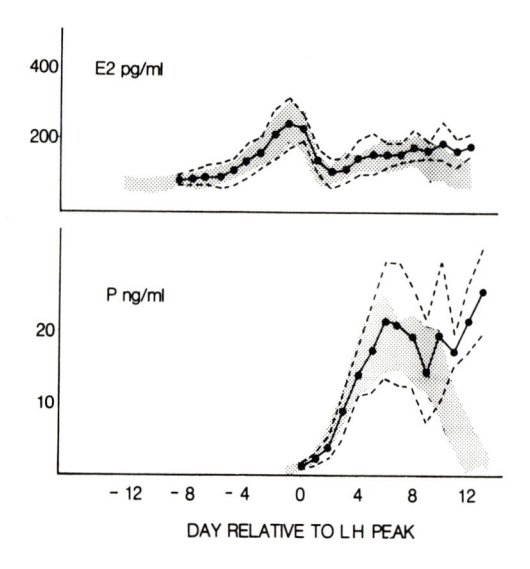

Figure 2. Mean (●) ± SD (– – –) concentrations of P and E_2 in spontaneous conception cycles (n = 8) compared with normal cycle ranges (hatched background) relative to the day of the LH peak. (For conversion to SI units, see Figure 1)

The major advance in the assessment of ovarian function that has occurred during the past few years is the application of ovarian ultrasonography to monitor follicular growth (Hackeloer et al., 1978). Figure 3 shows data from our laboratory (mean ± SD) for periovulatory follicular diameters (FD) and compares this with the profiles (mean ± SD) from the eight spontaneous conception cycles. No difference was observed between the ranges from the normal volunteers and the measurements determined during the conception cycles. In both normal volunteers (e.g. Figure 4) and in the conception cycles (Figure 3) the corpus luteum was no longer visible ultrasonically after day +6.

Figure 3. Mean (●) ± SD (– – –) follicular diameter (FD) relative to the day of the LH peak in spontaneous conception cycles (n = 8) compared with normal cycle ranges (hatched background). CL = corpus luteum

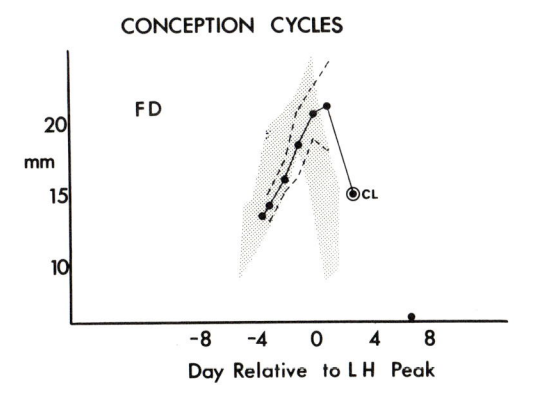

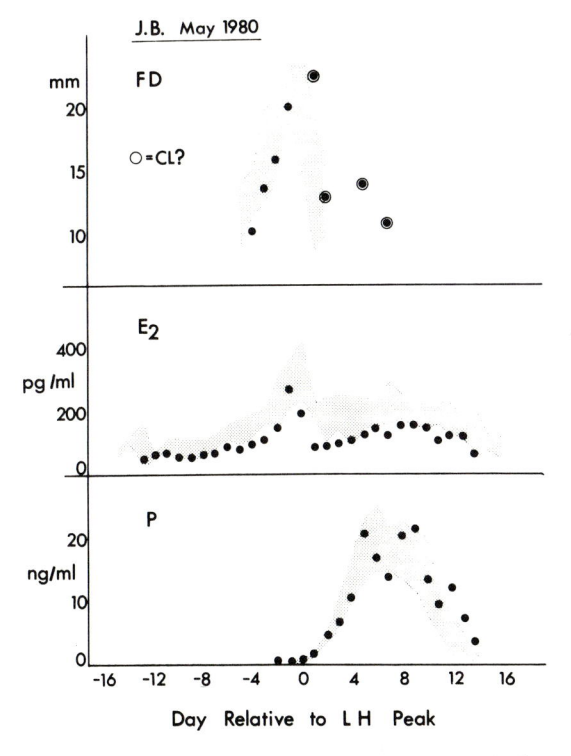

Figure 4. Concentrations of P and E_2 and follicular diameters (FD) in normal volunteer J.B. compared with normal cycle ranges (hatched background). (For conversion to SI units, see Figure 1.) CL = corpus luteum

We have therefore used these 'normal' ranges for steroid hormones and FDs for comparison with profiles from infertile women who might have abnormal ovarian function.

2.2 Hormone Profiles in Women with Unexplained Infertility

When the hormone profiles of the infertile women were compared with the 'normal' ranges, a number of endocrine anomalies could be described, viz. transient hyperprolactinaemia, short luteal phases, poor follicular maturation (PFM), poor P surge (PPS) and elevated basal LH (see Coutts *et al.*, 1978, for definitions). Although such anomalies were often found, 47% of the cycles from the infertile women had apparently normal hormonal profiles. Of the endocrine anomalies, PPS probably represents what many would regard as the deficient luteal phase. The PPS was diagnosed when P concentrations fell below the normal ranges on at least 3 of days +1 to +6. Thereafter during the remainder of the luteal phase P levels may remain subnormal or may rise into the 'normal' ranges. Figure 5 shows an example of the E_2 and P profiles of a

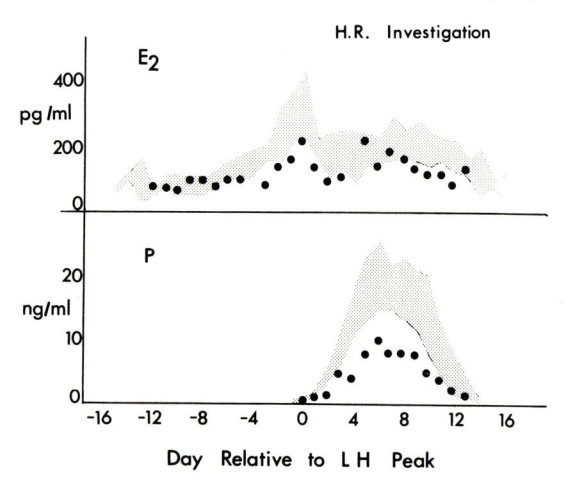

Figure 5. E_2 and P profiles in patient H.R. compared with normal cycle ranges (hatched background). (For conversion to SI units, see Figure 1)

patient with PPS who was deficient throughout the luteal phase. Figure 6 lists the incidence of these endocrine anomalies in 270 cycles from 247 women. There appears to be a loose association between elevated basal LH levels and PPS since many patients with the former anomaly also showed the latter.

Normal hormone profiles occurred in 47% of the cycles studied and, of the anomalies, PPS was the most common. This was observed in 38% of the cycles examined, and in 50% of these P levels rose later in the cycle to achieve normal concentrations during the mid- to late luteal phase (Figure 7). This finding

ENDOCRINE ASSESSMENT* OF CYCLES (n = 270)
FROM WOMEN WITH UNEXPLAINED INFERTILITY

ENDOCRINE ASSESSMENT	CYCLES	% OF CYCLES
NORMAL	128	47
TRANSIENT HYPERPROLACTINAEMIA	41	15
SHORT LUTEAL PHASE	15	6
POOR FOLLICULAR MATURATION (PFM)	24	9
POOR PROGESTERONE SURGE (PPS)	99¥	38
ELEVATED BASAL LH	32	12

* Based on daily plasma samples.
¥ 50 of these 99 cycles were deficient in P throughout.

Figure 6. Incidence of endocrine anomalies in menstrual cycles from women with unexplained infertility

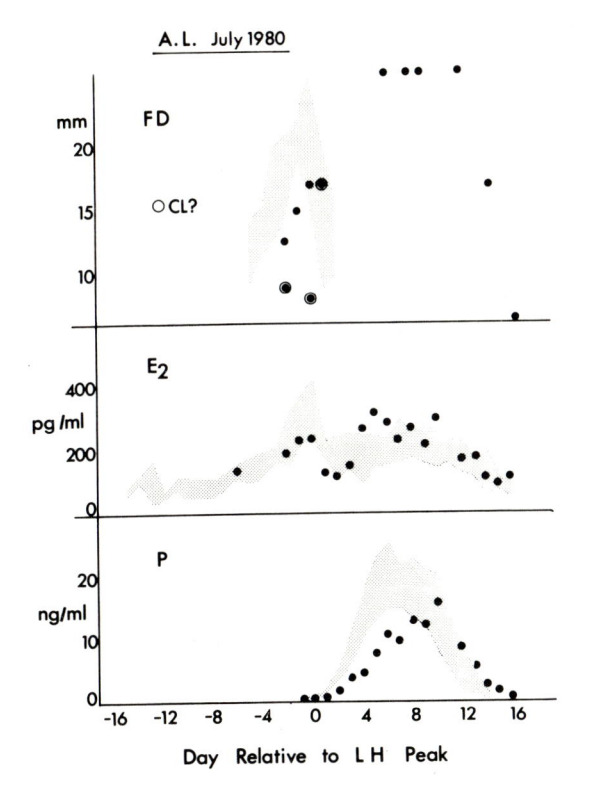

Figure 7. FD, E_2 and P profiles in a cycle from patient A.L. compared with normal cycle ranges (hatched background). (For conversion to SI units, see Figure 1.) CL = corpus luteum

underlines the statement in the Introduction that such deficiencies can only be correctly diagnosed where serial daily plasma samples are obtained.

2.3. Ovarian Ultrasound Profiles in Women with Unexplained Infertility

There are few reports of ovarian ultrasonography in the luteal phase and in general they indicate that the ovarian cystic structures can no longer be clearly visualized once luteinization is well established (Hackeloer et al., 1978) (Figures 3 and 4). However, when unexplained infertile patients were examined, a number of patients were shown to have retained luteal phase cysts (Coutts et al., 1982) (Figure 7) which appeared to represent failed rupture of the follicle and therefore possibly anovulation. These luteal cysts, which can grow to considerable sizes (30 to 50 mm in diameter), usually disappeared during the perimenstrual phase of the cycle (Figure 7). They were commonly found in association with PPS (Figures 7 and 8). A comparison of P indices in patients

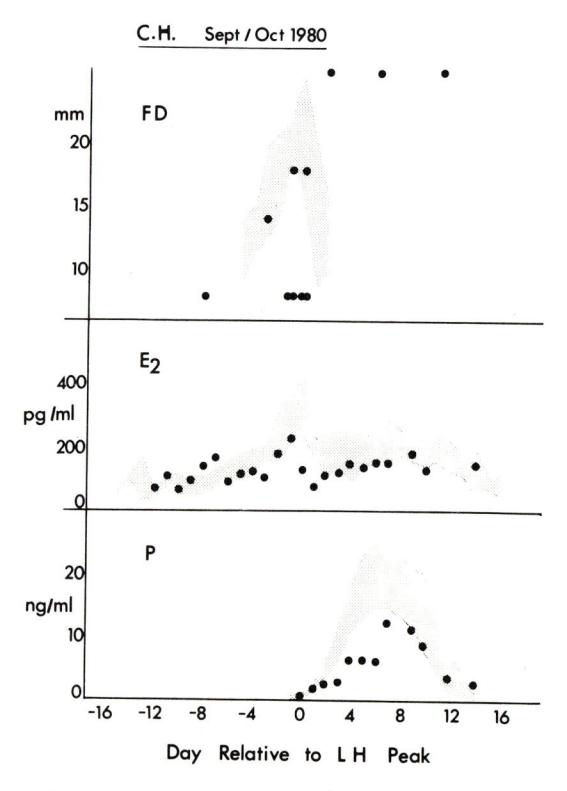

Figure 8. FD, E_2 and P profiles in a cycle from patient C.H. compared with normal cycle ranges (hatched background). (For conversion to SI units, see Figure 1)

who had retained luteal phase cysts ($n = 40$) with those who had normal luteal phase ultrasonography ($n = 100$) is shown in Figure 9. Although there was some overlap there was a correlation ($p < 0.01$) between subnormal P indices and the presence of luteal phase cysts.

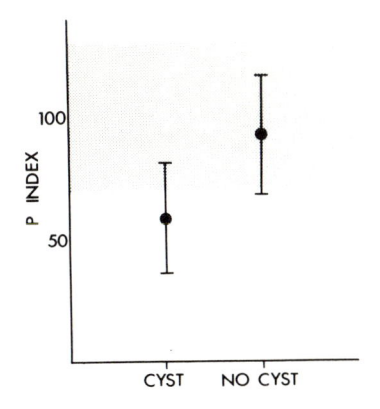

Figure 9. Comparison of P indices (mean ± SD) in patients with retained luteal phase cysts ($n = 40$) with those in patients showing normal luteal phase ultrasound ($n = 100$)

Some patients with unexplained infertility therefore showed deficient (abnormal) luteal phases as assessed by both endocrine hormone profiles (PPS) and ovarian ultrasonography (retained luteal phase cysts). Although the two diagnoses of luteal phase deficiency are not mutually exclusive, their regular coincidence indicates that at least in some patients the two deficiencies are related. Such abnormal luteal phases may be indicative of the luteinized unruptured follicle (LUF) syndrome (Koninckx et al., 1978).

Although abnormal luteal phases have been diagnosed as described above both endocrinologically and ultrasonically it is important to determine if they are a causative factor in unexplained infertility.

2.4. Treatment of Unexplained Infertility

Many treatment regimens have been attempted empirically for women with unexplained infertility including anti-oestrogen therapy (Fleming and Coutts, 1982), prolactin reducing agents (Craig et al., 1981) and gonadotrophin therapy (Black et al., 1981). None of these regimes produced significant improvements in terms of pregnancy rates. We have hypothesized that the lack of success of such regimens is due to the patients' functional pituitary–hypothalamic systems secreting endogenous gonadotrophins which compromise the various treatments. To alleviate this we have developed a treatment regimen involving pharmacological induction of hypogonadotrophic hypogonadism with a high

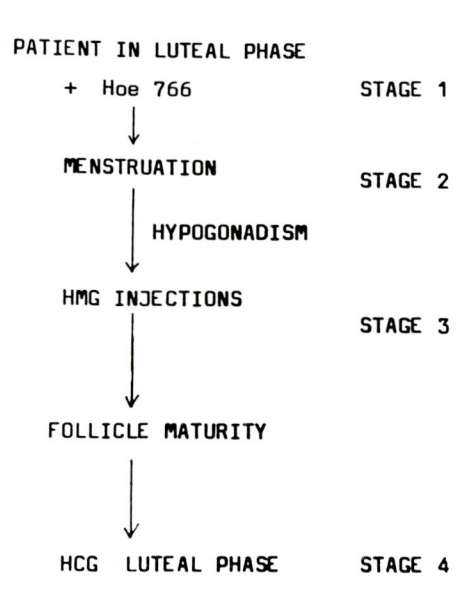

```
                    PROTOCOL


            PATIENT IN LUTEAL PHASE
                + Hoe 766                   STAGE 1
                       ↓
                  MENSTRUATION              STAGE 2
                       |
                       | HYPOGONADISM
                       ↓
               HMG INJECTIONS
                       |                    STAGE 3
                       |
                       ↓
               FOLLICLE MATURITY
                       |
                       |
                       ↓
             HCG   LUTEAL PHASE             STAGE 4
```

Figure 10. Outline protocol for therapy for women with unexplained infertility using a combined combination of HOE-766 and exogenous gonadotrophin. hMG = human menopausal gonadotrophin hCG = human chorionic gonadotrophin

dosage of the LHRH agonist D-Ser(tBu)^6-EA^{10}-LHRH; HOE-766, buserelin and concurrent induction of follicular growth and ovulation using exogenous gonadotrophins (human menopausal gonadotrophin, Pergonal, and human chorionic gonadotrophin). The outline of the protocol for this treatment (Fleming et al., 1982) is shown in Figure 10. After shutdown of the patient's own gonadotrophin release and consequently her ovarian activity (Figure 11), follicular growth and ovulation were induced (Figure 12). Five patients who showed PPS (abnormal luteal phases) in their investigation cycles (Figure 13) were treated initially. All of these patients who had a combined unexplained infertility of 51 years' duration conceived within seven treatment cycles each. These original five patients were investigated prior to the use of serial ovarian ultrasonography, but a further two patients who showed both PPS and retained luteal phase cysts have been treated successfully using this combined HOE-766 plus exogenous gonadotrophins regimen. To date, seven of ten patients with PPS have conceived on this therapy within a course of seven treatment cycles, whilst four patients who had normal endocrine profiles in their investigation cycles have failed to conceive within the same treatment period.

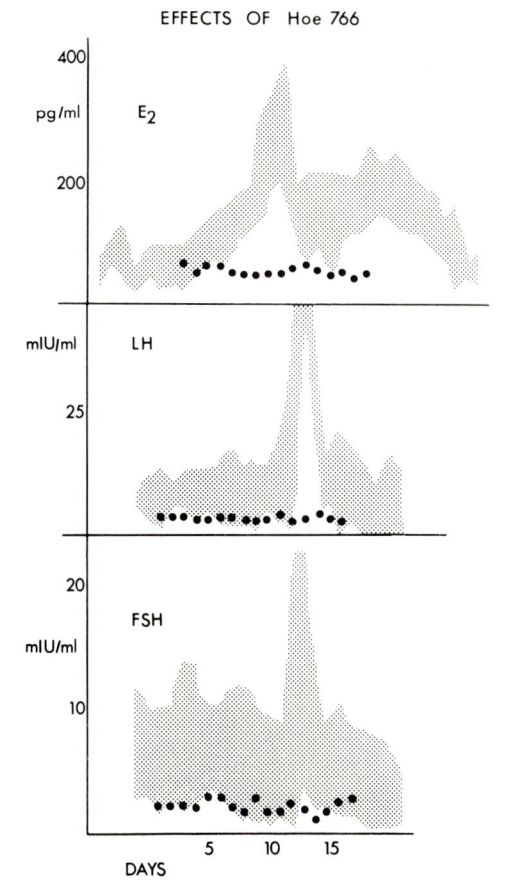

Figure 11. Effects of HOE-766 therapy on gonadotrophin and E_2 profiles in a woman with unexplained infertility. (For conversion to SI units, see Figure 1)

These preliminary results indicate that the abnormal luteal phase as diagnosed by PPS and retained luteal phase cysts is probably a causative factor in unexplained infertility, whereas in patients with unexplained infertility and normal luteal phase hormone profiles the cause of the infertility remains unexplained but, within the limitations of our present knowledge, this is probably unrelated to luteal phase endocrinology.

3. Conclusions

The abnormal luteal phase can be diagnosed both endocrinologically (PPS) and ultrasonically (retained luteal phase cysts), and in the series described in this chapter was found in 38% of cycles from women with unexplained infertility.

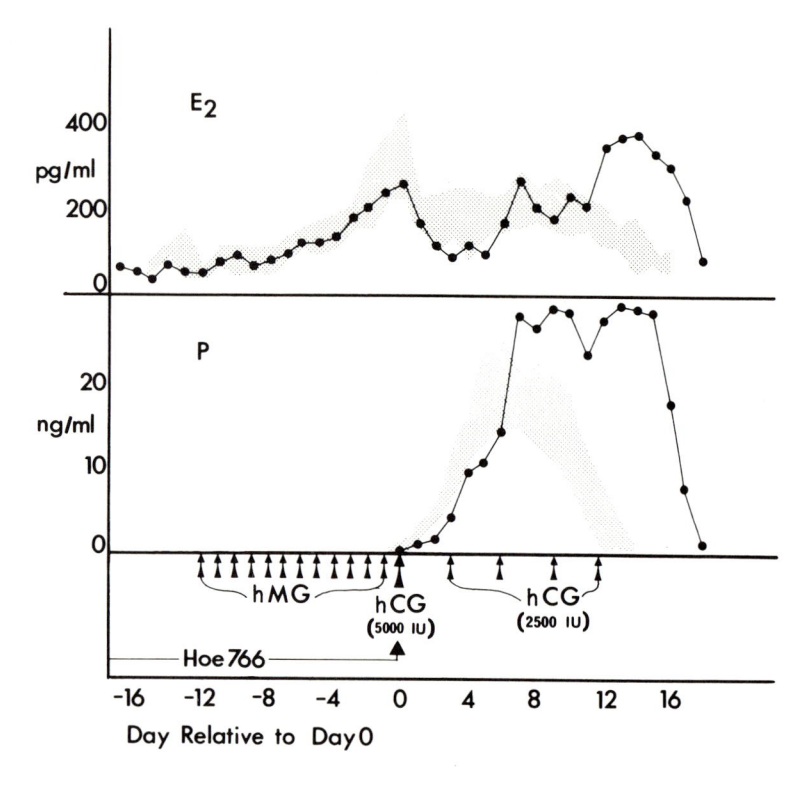

Figure 12. Induction of ovulation using exogenous gonadotrophins in a patient with unexplained infertility during simultaneous pituitary suppression with HOE-766. (For conversion to SI units, see Figure 1.) hCG = human chorionic gonadotrophin, hMG = human menopausal gonadotrophin

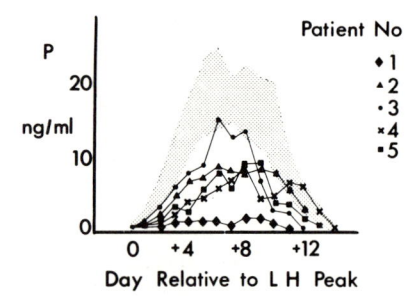

Figure 13. Pretreatment P profiles compared with normal cycle ranges (hatched background) in the first five patients treated with combined HOE-766 plus exogenous gonadotrophin therapy. (For conversion to SI units, see Figure 1)

Preliminary evidence, using a new treatment regimen, indicates that this luteal phase anomaly may be a causative factor in these patients' infertility. Pharmacological intervention in the patient with no apparent luteal phase abnormality is probably redundant.

Acknowledgements

The work presented in this chapter was done in collaboration with the following colleagues: R. Fleming, W.P. Black, D.H. Barlow, G.C. McCune, M.C. Macnaughton, A.H. Adam, V. Hood and M.P.R. Hamilton. Thanks are due to Mr. W. McNally for the art work. The HOE-766 was kindly donated by Hoechst UK Ltd. (Dr. P. Magill). Some of the work described was performed under a project grant (G82 00415 SB) from the MRC.

References

Adam, A.H. (1981). 'Infertile patients – ovarian ultrasound in relation to their management.' In *Ultrasound and Infertility* (ed. A.D. Christie), pp. 47–57. Chartwell-Bratt, Bromley, Kent.

Black, W.P., Fleming, R., Macnaughton, M.C., Craig, A., England, P., and Coutts, J.R.T. (1981). 'Infertility with normal menstrual rhythm; hormone profiles in response to HMG (Pergonal) treatment. In *Endocrinological Cancer: Ovarian Function and Disease*, (*Research on Steroids*) IX) (eds. H. Adlercreutz, R.D. Bulbrook, H.J. van der Molen, A. Vermeulen and F. Sciarra), pp. 370–373. Excerpta Medica, Amsterdam.

Coutts, J.R.T., Fleming, R., Carswell, W., Black, W.P., England, P., Craig, A., and Macnaughton, M.C. (1978). 'The defective luteal phase.' In *Advances in Gynaecological Endocrinology* (ed. H.S. Jacobs), pp. 65–91. Royal College of Obstetricians and Gynaecologists, London.

Coutts, J.R.T., Gaukroger, J.M., Kader, A.S., and Macnaughton, M.C. (1981). 'Steroidogenesis by the human Graafian follicle.' In *Functional Morphology of the Human Ovary* (ed. J.R.T. Coutts), pp. 53–72. MTP Press, Lancaster.

Coutts, J.R.T., Adam, A.H., and Fleming, R. (1982). 'The deficient luteal phase may represent an anovulatory cycle.' *Clin. Endocrinol.*, **17**, 389–394.

Craig, A., Fleming, R., Black, W.P., Macnaughton, M.C., England, P., and Coutts, J.R.T. (1981). 'Infertility with normal menstrual rhythm: hormone patterns before and during treatment with bromocriptine.' In *Endocrinological Cancer: Ovarian Function and Disease*, (*Research on Steroids* IX) (eds. H. Adlercreutz, R.D. Bulbrook, H.J. van der Molen, A. Vermeulen and F. Sciarra), pp. 293–298. Excerpta Medica, Amsterdam.

Fleming, R., and Coutts, J.R.T. (1982). 'Effects of clomiphene treatment on infertile women with normal menstrual rhythm.' *Br. J. Obstet. Gynaecol.*, **89**, 749–753.

Fleming, R., Adam, A.H., Barlow, D.H., Black, W.P., Macnaughton, M.C., and Coutts, J.R.T. (1982). 'A new systematic treatment for infertile women with abnormal hormone profiles.' *Br. J. Obstet. Gynaecol.*, **89**, 80–83.

Hackeloer, B.J., Fleming, R., Robinson, H.P., Adam, A.H., and Coutts, J.R.T. (1978). 'Correlation of ultrasonic and endocrinologic assessment of follicular development.' *Am. J. Obstet. Gynecol.*, **135**, 122–128.

Koninckx, P.R., Heyns, W., Corveleyn, P.A., and Brosens, I.A. (1978). 'Delayed onset of luteinisation as a cause of infertility.' *Fertil. Steril.,* **29,** 266–269.

Laufer, N., Mavot, D., and Schenker, J.G. (1982). 'The pattern of luteal phase plasma progesterone and estradiol in fertile cycles.' *Am. J. Obstet. Gynecol.,* **143,** 808–813.

Lenton, E.A., Sulaiman, R., Sobowale, O., and Cooke, I.D. (1982). 'The human menstrual cycle: plasma concentrations of prolactin, LH, FSH, oestradiol and progesterone in conceiving and non-conceiving women.' *J. Reprod. Fertil.,* **65,** 131–139.

The Luteal Phase
Edited by S.L. Jeffcoate
© 1985 John Wiley & Sons Ltd.

CHAPTER 8

Contraception in the luteal phase

M. G. ELDER
Institute of Obstetrics and Gynaecology,
Hammersmith Hospital,
Du Cane Road,
London, W12 0HS, UK

Contraception in the luteal phase means the use of a contraceptive method that allows complete or partial luteinization of the follicle and includes barrier methods, intrauterine devices and the safe period, none of which influence ovulation. Progestogen-only contraceptives inhibit ovulation in some cycles but sufficiently infrequently to allow them to be considered as a contraceptive method of the luteal phase. Progestogens released continuously from other delivery systems such as vaginal rings or subdermal implants may or may not inhibit ovulation, depending on the dose used. Finally, postcoital methods do not inhibit the luteal phase but affect implantation. Barrier methods, intrauterine devices and the safe period will not be considered in this paper.

1. Progestogen-only Oral Contraception

The progestogens used in this form of oral contraceptive are usually norethisterone and levonorgestrel. Failure rates vary considerably depending on the population studied and the nature of the trial, the range being between approximately 1.7 and 7.9 per 100 woman years (Hawkins and Elder, 1979).

1.1 Mode of Action

Progestogens have a variety of sites of contraceptive action which include the cervix, endometrium, fallopian tube and ovary. One of the modes of action of progestogen-only pills is the impairment of ovulation which occurs in approximately 50% of cycles. Luteal function may be impaired if ovulation has taken place, resulting in a diminished concentration of progesterone circulating for a reduced time, which is defined as an inadequate luteal phase. The synthetic progestogen will also compete with progesterone for receptor sites in the cells of target organs such as the fallopian tube and uterus. The tubal effects of progestogens are reduced ciliation and perhaps altered motility. The effects of norethisterone and norgestrel on the endometrium include increased stromal vascularity and oedema. The endometrial glands become atrophic and there is a reduction in mucus production. Endometrial biopsies taken during the second half of the cycle show that the progestogen changes the normal secretory pattern into an irregular secretory or inactive endometrium. This effect is dose dependent so that the higher the dose of progestogen used the fewer the biopsies that reveal a normal secretory phase (Landgren *et al.,* 1979). Clinical effects of this will be to cause irregular bleeding and to prevent implantation. This fact illustrates the major dilemma of this form of contraception which is that the higher the dose of progestogen used the more effective it will be as a contraceptive but the greater will be the bleeding irregularities.

The effects of progestogens on cervical mucus are to increase the protein content of the mucus and its sialic acid content. Sialic acid is part of the mucopolysaccharide component of the cross-linkages between glycoprotein fibrils of cervical mucus, and the greater its concentration the stronger are these cross-linkages. This causes an increased viscosity of the mucus. As a result of these changes there is a marked decrease in the spinnbarkeit, and in the ability of sperm to penetrate through this much more viscous mucus (Eckstein *et al.,* 1972).

A reason for the reduced efficiency of progestogen-only oral contraceptives is the difficulty of taking pills at a regular time each day to minimize the effect of the fluctuating blood levels.

In an attempt to reduce the total amount of steroid ingested and the changing blood levels of progestogen that occur with oral or intramuscular administration as well as to avoid the first-pass effect through the liver which occurs with oral contraceptives, new delivery systems releasing small amounts of oral progestogens at a constant rate are being developed. Zero-order release is defined as a constant release rate of a substance into the body over a period of time which is independent of the concentration of the substance remaining in the matrix of the delivery system. A subdermal implant and an intravaginal ring to provide zero-order release of a progestogen are being investigated.

1.2. *Subdermal Implant*

Release of a drug from an implant that is not eroding depends on the following criteria: (1) diffusion of the drug to the inner surface of the polymer barrier; (2) dissolution of the drug into the polymer and diffusion across the polymer barrier; (3) desorption from the barrier surface and dissolution of the drug into the surrounding tissue fluid.

The new subdermal implant being investigated is a single one, which is 2.5 cm long and 0.3 cm in diameter. The capsule is polyepsilon caprolactone, which will degrade extremely slowly to hydroxycaproic acid. Levonorgestrel is in suspension with ethyl oleate within the polymer capsule and is released at the rate of approximately 45 μg per day. A reasonably constant release rate of levonorgestrel was obtained during the 30-day implant period. No local irritation was observed. The side-effects were intermenstrual spotting in two subjects and slight mastalgia in two subjects. Further studies are proceeding with this small single implant system which could last for up to 2 years. Whether it will biodegrade effectively or have to be removed has to be determined.

A larger silastic device called Norplant, which releases approximately 40 μg of levonorgestrel daily, has recently undergone extensive clinical trials. The device is effective for up to 4 years and possibly longer but has to be removed. There is constant release of levonorgestrel from such a device, showing a linear relationship between the total amount of levonorgestrel released (in mg) and the duration of its insertion (in hundreds of days). Cycle control with this system seems to be quite good as over 40% of women have between 11 and 12 episodes of bleeding during a year (Diaz *et al.*, 1979).

1.3. Intravaginal Ring

The intravaginal ring is a silicone rubber vaginal device approximately 60 mm in diameter, the diameter of the core being 9 mm. The types of vaginal ring releasing progestogen that have been developed are as follows: (A) a ring which is entirely impregnated with the progestogen; (B) a large central core containing progestogen covered by a thin outer layer of inert silastic; (C) a samll inner core containing progestogen covered with a thick outer layer of inert silastic; and (D) a sandwich which has an innermost layer of silastic, then a steroid impregnated layer, and finally an outer layer of silastic (Figure 1). Rings have been developed to release large doses of progestogen with or without oestrogen which will inhibit ovulation. They are inserted for 3 weeks and then removed for the fourth week when withdrawal bleeding occurs (Sivin *et al.*, 1981).

An alternative philosophy is to develop a vaginal ring which is meant to be retained for up to 3 months or possibly longer and which releases constant but low doses of progestogen which will not inhibit ovulation. Nearly constant release of levonorgestrel at 20 μg per 24 h can be obtained from such a vaginal device for up to 3 months. Plasma norgestrel levels reach a maximum level by 4 h and the removal half-life of levonorgestrel in subjects using this vaginal delivery system is approximately 16 h (Landgren *et al.*, 1982). These data suggest that the levonorgestrel-releasing vaginal ring will be effective soon after insertion and that contraceptive levels of the steroid will still be circulating some hours after removal of the device.

Low-dose progestogens released from intravaginal rings have a varying effect

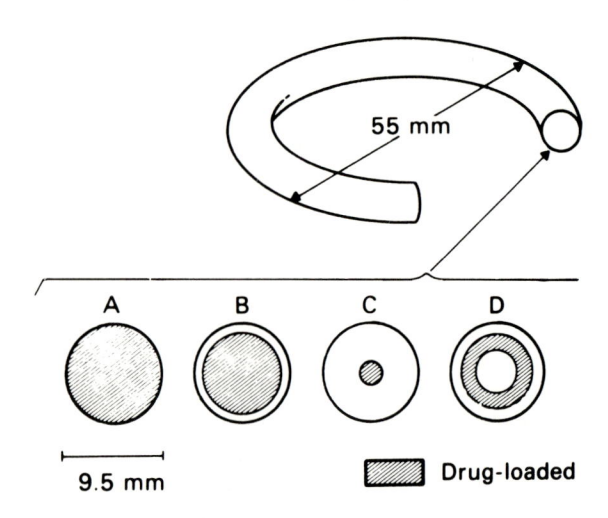

Figure 1. Vaginal ring with different types of progestogen loading. A. Entire impregnation. B. Central core of progestogen with outer inert layer. C. Similar to B with small central core and thicker outer layer. D. A sandwich with a progestogen layer between inner and outer inert layers

on ovulation. An intravaginal ring releasing 200 µg of norethisterone is moderately effective at preventing pregnancy but gives poor cycle control because 80% of the subjects had anovulatory cycles or inadequate luteal function, both of which are associated with irregular menstruation. This compares with a total of only 18% of subjects with anovulatory cycles or inadequate luteal function resulting from use of the vaginal ring releasing 50 µg of norethisterone. However, this dose resulted in a number of pregnancies.

Fifty-two per cent of patients using the vaginal ring releasing 20 µg of levonorgestrel showed a normal ovulatory-like pattern, whereas 48% of subjects have anovulatory cycles or cycles with an inadequate luteal function (Landgren et al., 1982). It is likely that cycle control will be changed in many users but will be within acceptable limits for 85% of subjects.

A multicentre clinical trial of these intravaginal rings releasing 20 µg of levonorgestrel is being carried out, and preliminary data suggest a pregnancy rate of 3.1 calculated by life-table analysis at 360 days. The rate of removal for both medical and non-medical reasons as well as the continuation rate are similar to those of many other contraceptives studied in clinical trial. The principal medical reason for removal is bleeding problems, but this may turn out to be less of a problem than with oral progestogen-only contraception when plasma levels of the progestogen are variable. These data suggest that the intravaginal ring might have considerable appeal and could give acceptable results in terms of use effectiveness and incidence of side-effects.

Progestogens used in oral contraceptives have been shown to alter plasma lipids and are thought to be the major factor responsible for arterial disease among subjects using combined oral contraceptives. Recent studies of the use of intramuscular progestogens such as norethisterone enanthate have shown that there is a 25% reduction in high-density lipoprotein (HDL) cholesterol (Fotherby *et al.*, 1982). A recent study of plasma lipid levels following the use of high doses of norgestrel and oestradiol released from an intravaginal ring has also shown a 25% reduction in HDL cholesterol (Ahren *et al.*, 1982). The concept of a continuous release of a low dose of progestogen should avoid or minimize such an alteration. A study of 15 women in whom plasma lipid levels were measured before and 12 weeks after the insertion of an intravaginal ring releasing 20 μg of levonorgestrel shows that there is no effect on mean plasma triglyceride levels or cholesterol levels. Additionally there is no difference in the ratio of cholesterol levels found in the high-density, low-density plus very-low-density cholesterol lipoprotein fractions to high-density cholesterol, the so-called atherogenic ratio (Elder *et al.*, 1984). These data suggest that the method has no short-term effect on lipids and therefore, unlike other steroidal contraceptive regimens (progestogen alone or combined with oestrogen) it might be free from the additional risk of arterial disease. Before drawing firm conclusions, long-term metabolic studies will need to be carried out and epidemiological data collected. There was no difference in the oral glucose tolerance tests carried out before and after 12 weeks of use of this intravaginal ring (Elder *et al.*, 1984). These preliminary data are encouraging in terms of the possible metabolic safety of this contraceptive.

2. Postcoital Contraception

Let us now consider postcoital contraception. This has been administered in three ways: (1) the administration of oestrogens or combined oestrogen–progestogen pills after intercourse; (2) the insertion of an intrauterine device within 36 h of intercourse; and (3) the administration of danazol. All methods prevent implantation. The steroids used are high doses of stilboestrol, 50 mg per day in 4–5 divided doses, or ethinyloestradiol 2–5 mg per day in 3–5 divided doses. These large doses of oestrogens cause side-effects such as nausea and vomiting. They are effective postcoital contraceptives in emergency cases but they are unacceptable in routine use because of their side-effects. For the older woman who may have intercourse less frequently, the administration of a progestogen postcoitally on a more regular basis has been tried. Levonorgestrel in a dose of 150 μg taken approximately eight times per month postcoitally results in an extremely high failure rate. However, increasing the dose of levonorgestrel to 300 μg postcoitally with the same frequency of intercourse results in an acceptable failure rate of approximately 4.9 per 100 woman years (Kesseru *et al.*, 1973).

Danazol, administered in a dose of 400 mg twice daily, has been investigated as a possible postcoital agent. It has been compared with the use of combined ethinyloestradiol (30 μg) and levonorestrel (250 μg) administered twice during a period of 24 h postcoitally and with the postcoital insertion of an intrauterine device (Guillebaud et al., 1983). A double dose of the combined oral contraceptive pill leads to a moderately high incidence of nausea and vomiting while danazol causes very little nausea. The incidence of other side-effects such as headache is also higher with the use of the combined oral contraceptive pill. However, as both compounds were being used in an emergency fashion rather than as a routine contraceptive, the high incidence of side-effects should not be overstressed. Rather, the greater effectiveness should be the more important factor in deciding upon an appropriate regimen.

What is the mode of action of these two postcoital regimens? To try and answer this question a study was set up in which a number of subjects had endometrial biopsies carried out during treatment and control cycles (Guillebaud et al., 1983). Daily urinary estimations of luteinizing hormone (LH) were carried out from day 10 onwards and all biopsies were taken on the third day after the LH peak. Cytosol and nuclear oestrogen receptor, cytosol progesterone receptor and isocitrate dehydrogenase estimations were carried out on the endometrial tissue. Administration of the combined pill led to a marked reduction in both nuclear and cytosol oestrogen receptor and cytosol progesterone receptor estimated on day 3 after the LH peak. There is a marked increase in the activity of isocitrate dehydrogenase (Figure 2).

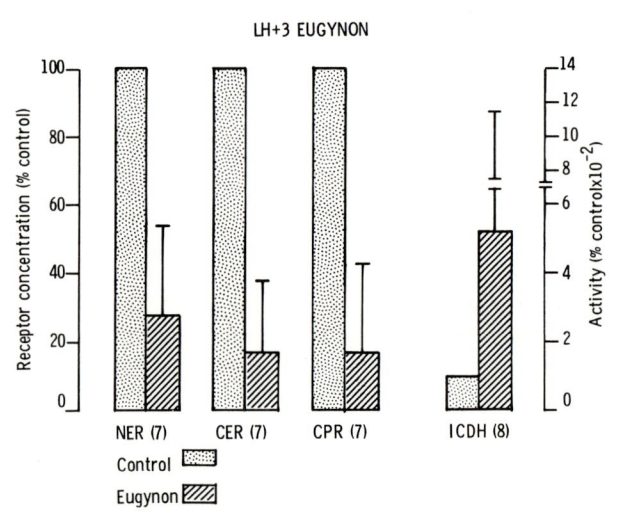

Figure 2. Effect of postcoital Eugynon on levels of nuclear oestrogen receptor (NER), cytosol oestrogen receptor (CER), cytosol progesterone receptor (CPR) and isocitric dehydrogenase activity (ICDH). Samples were taken on day 3 after the LH peak and values are expressed as a percentage of controls

Levonorgestrel and ethinyloestradiol will bind to progesterone and oestrogen receptor, respectively, thereby reducing their further production. Inhibition of cytosol and nuclear oestrogen receptor will reduce the synthesis of progesterone receptor. Suppression of receptor synthesis with an increase in isocitrate dehydrogenase, an enzymatic index of progesterone activity, suggests that the combined oral contraceptive is acting at the endometrial receptor level. Danazol causes a smaller reduction in both cytosol and nuclear oestrogen receptor and cytosol progesterone receptor. Danazol binds non-specifically to a number of steroid receptors and does not alter isocitrate dehydrogenase activity. Its mode of action is at present not clear.

In conclusion, steroidal contraception without inhibition of ovulation and with minimal metabolic effects has considerable advantages provided a suitable delivery system can be developed which will allow a controlled release of sufficient progestogen to prevent pregnancy and still produce an acceptable bleeding pattern. When more is know about the biochemical causes and possible control of irregular bleeding in these women, then the scope for this type of contraception could be considerable.

References

Ahren, T., Lithell, H., Victor, A., Vessby, B., and Johansson, E.D.B. (1982). 'Serum lipoprotein and apolipoprotein changes during treatment with a contraceptive vaginal ring containing levonorgestrel and oestradiol.' *Acta Obstet. Gynecol. Scand.*, **61**, 400–504.

Diaz, S., Pavez, M., Robertson, D.N., and Croxatto, H.B. (1979). 'A three year clinical trial with levonorgestrel silastic implants.' *Contraception*, **19**, 557–573.

Eckstein, P., Whitby, M., Fotherby, K., Butler, C., Mukherjee, T.K., Burnett, J.B.C., Richards, D.J., and Whitehead, T.P. (1972). 'Clinical and laboratory findings in a trial of norgestrel, a low-dose progestogen only contraceptive.' *Br. Med. J.*, **iii**, 195–200.

Elder, M.G., Lawson, J., Lucas, S., Hamawi, A., and Fotherby, K. (1984). 'Effect of a contraceptive vaginal ring releasing 20 µg levonorgestrel daily on blood lipid levels and glucose tolerance.' *Contraception* (in press).

Fotherby, K., Trayner, I., Howard, G., Hamawi, A., and Elder, M.G. (1982). 'Effect of injectable norethisterone enanthate on blood lipid levels.' *Contraception*, **25**, 435–446.

Guillebaud, J., Kubba, A., Rowlands, S., White, J., and Elder, M.G. (1983). 'Post-coital contraception with danazol, compared with an ethinyl oestradiol – norgestrel combination or insertion of an intrauterine device.' *J. Obstet. Gynaecol.*, **3**, 864–868.

Hawkins, D.F., and Elder, M.G. (1979). *Human Fertility Control*, p. 102. Butterworth, London.

Kesseru, E., Larranaga, A., and Parada, J. (1973). 'Post-coital contraception with d-norgestrel.' *Contraception*, **7**, 367.

Landgren, B.M., Johannisson, E., Masironi, B., and Diczfalusy, E. (1979). 'Pharmacokinetic and pharmacodynamic effect of small doses of norethisterone released from vaginal rings continuously during 90 days.' *Contraception*, **19**, 253–271.

Landgren, B.M., Johannisson, E., Masironi, B., and Diczfalusy, E. (1982). 'Pharmacokinetic and pharmacodynamic investigations with vaginal devices releasing levonorgestrel at a constant, near zero order rate.' *Contraception*, **26**, 567–585.

Sivin, I., Mishell, D.R., Victor, A., Diaz, S., Alvarez-Sanchez, F., Neilson, N., Akinla, O., Pyorala, T., Coutinho, E., Faundes, A., Roy, S., Brenner, P., Ahren, T., Pavez, M., Brache, V., Giwa-Osagie, O.F., Fasan, M.O., Zausner-Guelman, B., Darze, E., da Silva, J.C.G., Diaz, J., Jackanicz, T.M., Stern, J., and Nash, H.A. (1981). 'A multicentre study of levonorgestrel–estradiol contraceptive vaginal rings. I. Use effectiveness.' *Contraception*, **24**, 341–358.

Index

Abnormal luteal phase, 101–113, 118
  and infertility, 62, 70–80, 90,
    101–113, 116–118
  progesterone levels, 90, 101–103, 116
  prolactin and, 76–82
Adenyl cyclase
  LH dependence, 75
  prostaglandins and, 26–27, 32
Alkaline phosphatase, 51
Androgen
  aromatization of, 4, 11, 17–18, 26,
    34, 74
  in follicular fluid, 6–10
  in theca cells, 8–9
  synthesis, 8–13
Androstenedione, 6, 10, 13, 28
Anti-oestrogens, 89
Aromatase, 2, 4, 5, 8–11, 18, 26, 34, 74
Aromatization, 4, 8–11, 16–18, 26, 34, 74
Atherogenic ratio, 119

Baboons, 30
Basal body temperature, 101
Blastocyst, 43
  implantation of, 25, 47–48, 50–51,
    57–58, 95
Blood–follicle barrier, 12–14
Bromocriptine, 36, 73–74, 76, 80, 89

Cervical mucus, 116
Cholesterol, 9, 12–13, 28
  ester hydrolase, 12, 75
  ester synthetase, 75
  side-chain cleavage, 18
Clomiphene (citrate), 89

Closure reaction, 47
Conceptional cycle
  luteal function, 62–64, 90–96,
    104–105
  progesterone levels, 90–96, 104–105
Contraception in luteal phase, 115–121
  post coital, 119–121
Corpus albicans, 73
Corpus luteum
  formation and maintenance, 11–19,
    25–30
  gonadotrophic support, 14–19,
    26–30, 95–99
  in pregnancy, 61–62, 72, 76
  interaction with embryo, 61–68
  life span, 30, 32, 63, 74
  luteolysis, 25–39
  morphology, 27
  prolactin and, 26, 30, 71–82
  prostaglandin production, 32–33
  regression of, 30–31
  uterus and, 43–58
  vascular control, 11–14
Cow, 76
  luteolysis, 31–33
  prolactin receptors, 72
Cumulus expansion, 6
Cyclic AMP, 4, 18
Cynomolgus monkey, 36
Cytochrome P450, 28

Danazol, 119–121
Decidua, 52
Decidualization, 51
Deficient luteal phase, 62, 76, 79–80,
  90, 101–113, 116–118

Dehydroepiandrosterone, 28
20α-Dihydroprogesterone, 75
Dopamine agonists, 36, 73–74, 76, 80, 89

Early embryonic mortality, 67–68
Early pregnancy factor, 64
Embryo
  interaction with corpus luteum, 61–68
Embryo transfer, 50
Embryonic signals, 63–64
Endometrium
  morphology, 44–58
  ovarian hormones and, 43
  prostaglandins in, 34

Fecundability (in man), 67–68
Follicle, 1–12
  diameter, 104–109
Follicle-stimulating hormone (FSH)
  actions, 3–4, 9–10, 15–19, 30, 72, 74
  in blood, 93–94, 102
  receptors, 15, 18, 26
Follicular fluid
  androgens, 6, 9
  androstenedione, 6, 10
  lipoprotein in, 13–15
  oestradiol in, 2–3, 6, 9–11
  plasma proteins in, 14–15
  progesterone in, 2–3, 6, 8

Glands, uterine, 44, 50–51
Gonadotrophins
  episodic secretion of, 29–30
Gonadotrophin-releasing hormone (*see* Luteinizing Hormone-Releasing Hormone)
Granulated (K) cells, 27, 51–52
Granulosa cells, 2–3
  aromatase in, 2, 4–5, 8–11, 26, 34
  in vitro studies, 5–11, 15–17, 26, 32–33, 78–80
  luteinization of, 7, 78–79
Granulosa-lutein cells, 11–15, 18, 27
Guinea Pig
  implantation in, 49–50
  luteolysis in, 31–33

High-density lipoprotein (HDL) 12–15, 119

Human chorionic gonadtrophin (hCG), 64–66
  administration, 89–95, 110–112
  antiserum, 29–30, 65–66
  binding to corpus luteum, 30–31
  biological action, 7–11, 37–38, 65–66, 78–79
  half-life, 28, 95
  immunology, 65
  measurement of, 65–66
  receptor binding, 66
  secretion of, 25
  structure, 64–65
Human menopausal gonadotrophins (hMG), 89–95, 109–111
Human placental lactogen, 66
17-Hydroxyprogesterone, 28, 63
20α-Hydroxysteroid dehydrogenase, 75–76,
3β-Hydroxysteroid dehydrogenase, 26, 28, 36
Hyperprolactinaemia, 79–82, 107
Hypophysectomy, 75–77
Hypoprolactinaemia, 78
Hysterectomy, 75–77
  luteolysis after, 31–32

Implantation, 25
  in mouse, 45–58
  species variation, 44–49
Indomethacin, 34
Infertility
  and deficient luteal phase, 62, 79–80, 90, 101–113, 116–118
  hyperprolactinaemia, 79–82
  lactational, 80–82
  treatment of, 89–99
  unexplained, 101–111
Intrauterine device, 119–121
Intravaginal rings
  for contraception, 117–119
Isocitrate dehydrogenase, 119–121

K cells, 27, 51–52

Lactational infertility, 80–82
LDL-cholesterol, 12–15, 28
Levonorgestrel
  intravaginal ring, 117–119
  oral, 115–116
  subdermal implant, 116–117

INDEX

Lipoprotein
　high density (HDL), 12–15
　in follicular fluid, 13–15
　low density (LDL), 12–15, 28
Low-density lipoproteins, 12–15, 28
　receptors, 12, 18, 26
Luteal cells
　in vitro, 78–79
Luteal cyst, 108–109
Luteectomy, 62
Luteinized unruptured follicle (LUF), 14–15, 109
Luteinizing hormone
　actions, 3–19, 26–30, 75
　episodic secretion, 30, 80–82, 96–99
　plasma levels, 1–2, 30–31, 102
　receptor, 3–4, 18, 26, 28, 75–76
　surge of, 1–6, 8–10, 17, 26, 72–73, 80
Luteinizing Hormone-Releasing Hormone (LHRH)
　analogues, 36–37, 110
　-like peptide, 36–38
　pulsatile therapy, 89–99
　receptors, 26, 37
Luteolysis, 63–64
　control of, 25–39
Luteolytic agents, 30–31

Marmosets, 30
Menstruation, 54–56
Mesometrium, 44
Metrial glands, 44, 50–51
Mitosis
　in uterus, 46–47
Mouse uterus, 45–58
Myometrium
　ovarian hormones and, 43

Norethisterone, 115, 117–118
Norplant, 117

Oestradiol
　in corpus luteum, 28, 34–36
　in follicular fluid, 2–3, 6, 9–11
　in pregnancy, 63
　plasma concentrations, 2, 72, 90, 102–108
　receptor, 75
　synthesis, 8–11

Oestrogen
　action on uterus, 45–50
　luteolytic effect, 9, 34–36
　post coital contraception, 119–121
　receptors, 119–121
Oestrone, 28, 63
Oocyte, 6–12
Ovulation, 6, 8
　induction of, 28–29, 89–99
Oxytocin, 37–38

Parturition, 43
Post-coital contraception, 119–121
　danazol, 119–121
　intrauterine device, 119–121
　oestrogens, 119–121
Pregnancy, 43
　corpus luteum, 61, 68, 72, 76
　tests, 65
Pregnenolone, 18, 28
　conversion to progesterone, 18, 26
Progesterone
　action on uterus, 46–58
　in abnormal luteal phase, 90, 101–113, 116
　in conception cycles, 90–96, 104–105
　in follicular fluid, 2–3, 6, 8
　index, 102, 108–109
　prolactin and, 26, 74–82
　secretion, 6, 11–14, 25, 28, 62, 74–82, 90–97
　surge, 106–109
　synthesis, 4, 7–8, 33
Progestogen-only contraception, 115–119
Prolactin
　and corpus luteum, 26, 30, 71–82
　and progesterone secretion, 26, 74–82
　in blood, 72–74, 78–82, 102, 107
　in pig, 72, 76–78
　in primates, 72, 76–77
　in rat, 72, 75
　in rhesus monkey, 72
　in sheep, 76–78
　receptors, 4, 30, 71–75
Prostaglandin
　effect on adenyl cyclase, 26–27, 32
　in endometrium, 34
　luteolysis and, 31–33
　synthesis, 6

Prostaglandin E_2, 26, 33
Prostaglandin $F_{2\alpha}$, 25–26
   luteolytic action, 31–33, 63
   oxytocin and 37–38
   receptors for, 26, 32
Proteoglycans, 4
Pulsatile infusion
   LHRH, 89–99
   pump, 90
Radioreceptor assays, 66
Rat
   oestrus cycle of, 73–74
   prolactin receptors, 72
Receptors
   FSH, 15, 18, 26
   hCG, 66
   LDL, 12, 18, 26
   LH, 3–4, 18, 26, 28, 75–76
   LHRH, 26
   oestrogen, 75, 119–121
   progesterone, 120–121
   prolactin, 4, 30, 71–73
   prostaglandins, 26, 32
Relaxin, 37–38
Rhesus monkey, 28, 32–33

Sheep, 76–78
   luteolysis in, 31–33, 63
Subdermal implants, 115

Suckling, 80–82
Sulpiride, 89
Syncytiotrophoblast, 64

Testosterone, 18, 28
Theca cell, 7–8
   androgen synthesis, 8–13
Theca interna, 17
Theca lutein cells, 13, 27
Triploidy, 51
Trophoblast, 49, 64

Ultrasonography
   of ovary, 90–96, 101–109
   of uterus, 94–95, 98
Uterus
   closure reaction, 47
   epithelial cells, 47–50
   glands, 44, 50–51
   lumen of, 47, 54
   microvilli, 47–48
   mitosis in, 46–47
   mouse, 45–58
   progesterone action on, 46, 58
   prostaglandin release, 31–33
   response to corpus luteum, 43–58
   stroma, 44, 50–58
   ultrasonography of, 94–95, 98